CONQUERI

MATH PHOBIA

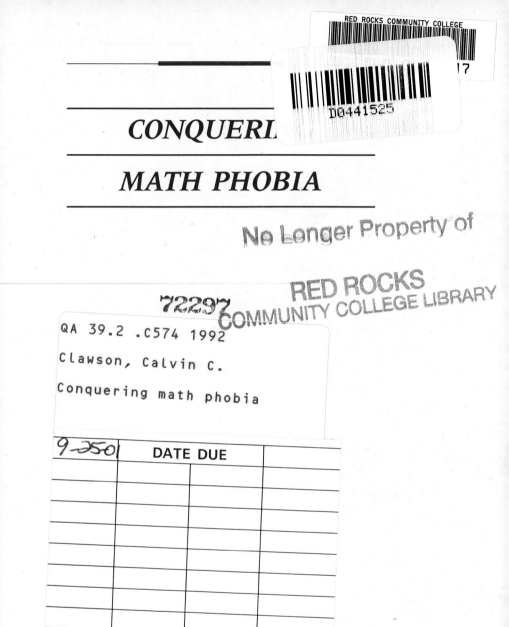

CONQUERING

MATH PHOBIA

A Painless Primer

Calvin C. Clawson

John Wiley & Sons, Inc.

New York · Chichester · Brisbane · Toronto · Singapore

Recognizing the importance of preserving what has been written, it is a policy of John Wiley & Sons, Inc., to have books of enduring value published in the United States printed on acid-free paper, and we exert our best efforts to that end.

Clawson, Calvin C.
    Conquering math phobia : a painless primer / Calvin C. Clawson.
        p.   cm.
    Includes bibliographical references.
    ISBN 0-471-52898-6 (pbk. : alk. paper) : $14.95 (est.)
    1. Mathematics.   I. Title.
QA39.2.C574   1992
510—dc20
                                            91-41249

    Printed in the United States of America

    10 9 8 7

This book is dedicated to Susan Bara Clawson
for unwavering faith and perseverance.

ACKNOWLEDGMENTS

Many people helped make this book a reality. Many thanks to Ralph Penner for carefully reshaping the manuscript into a useful tool for learning. Thanks to Steve Ross of John Wiley & Sons for shepherding this project with gentleness and wisdom. Thanks must also go to Jeff Herman and Deborah Adams for first recognizing the need for this book. Thanks go to my workshop friends who enthusiastically reviewed the manuscript; Marie Edwards, Bruce Taylor, Linda Shepherd, Phyllis Lambert, and Brian Herbert. Thanks to Dr. Robert Casad of Green River Community College for his constant support and encouragement.

Finally, I want to thank my many professors at the University of Utah for their patience, talent, and love of mathematics, which made this book possible.

CONTENTS

CHAPTER 1

WHAT IS

MATH PHOBIA?

Have you ever found yourself in situations like these?

- You're in the checkout line at the supermarket. The clerk takes your money and gives you change. You put the change in your wallet or purse without making sure that the change is correct because you're not sure how to do that.
- You're in a restaurant dining with friends. When the bill arrives, one of your friends asks you to split up the bill and tell each person how much to pay. You decline the task, trying to say something gracious, because you don't know how to divide and you're not sure how to figure out the size of the tip.
- You sit at the kitchen table, checks and bills spread out before you, ready to tear your hair out because you cannot balance your checking account.

If you have ever been in any of these situations, you are definitely not alone. You probably suffer from math phobia, or fear of numbers. You are not necessarily math-illiterate, but you are afraid of dealing with numbers. Millions of other people share the same fear.

Math phobia is neither a disease nor an affliction. However, it is a handicap that can have a dramatic impact on your life. As a mathophobe you will be hampered in your ability to perform calculations, and, unfortunately, you cannot simply ignore mathematics. At supercomputers in windowless rooms across the country, men and women perform simple computations that determine the amount of your mortgage payment, how much you pay for groceries, and the SAT scores of your sons or daughters. If you cannot perform these same computations, you are effectively allowing others to run your life.

The academic weakness of our schools cuts across subjects, but students seem to be especially sensitive to failures in learning mathematics. Difficulties in English or geography in early grades are often corrected in later grades. With mathematics, however, students who fall behind experience extreme difficulty in trying to catch up to their expected level of performance. In their book, *Mind Over Math*, Dr. Stanley Kogelman and Dr. Joseph Warren report that, although some students develop a dislike for math in elementary school, students most commonly have negative experiences with math between the seventh and tenth grades.[1]

As a mathophobe, you probably experience one or more of the following emotional responses whenever you deal with math:

- Sweaty hands
- Palpitating heart
- Loss of concentration
- A general sense of uneasiness
- Mild stomach pains or cramps
- Muscles tightening

You may continue to have these reactions until you have replaced your negative mathematical experiences with a set of positive experiences.

This book can help you not only to overcome your math phobia, but also to see the power, beauty, and elegance of mathematics. It can help you to change your attitude toward math and to propel you toward becoming math-competent. You don't need expensive cassettes to put under your pillow, special schools, or high-priced programs. All you need is a commitment to your future and a resolve that you will no longer be a victim of your fear of numbers. The first step is to destroy six common myths about math.

[1]Dr. Stanley Kogelman and Dr. Joseph Warren, *Mind Over Math*, (New York: McGraw-Hill, 1978): 16.

Myth One
Math is boring.

The best way to destroy this myth is to remember that nothing is inherently boring. Think of the old saw, "One man's meat is another man's poison." Some people read novels for pleasure; others haven't read a novel in years. But as the avid novel reader knows, people who have never acquired a taste for novels have probably never read the kinds of stories that are most likely to appeal to them. It's the same with mathematics: If you think math is boring, it's a safe bet that you've had only boring math teachers.

But boring math teachers have an advantage over their bored students because of their math skills. Consider the following: Young Lord Donald, the son of the richest man on earth, had a boring math teacher of his very own. One day she said to him, "Now if a train leaves Point A at 12:00 noon and travels at a speed of 55 miles per hour ... "

"Ugh!" grimaced young Lord Donald. "This is the most boring math question in the world! I *hate* this question! I'll tell you what: There are only 30 days left in this school year. I'll give you $5,000 for each day you don't ask this question."

"No way, Donald. I'm a boring math teacher. I have a mission. What do you take me for, anyway?"

"$10,000 a day."

"It's a deal."

"All riiiiight!" Young Lord Donald took out his calculator: 10,000 dollars per day × 30 days = $300,000.

"Unless, of course, you'd rather start with a penny and double it from there," said the boring math teacher.

"A penny?"

"Sure. You give me a penny today, two cents tomorrow, four cents the next day, and so on, for 30 days."

Donald started trying to figure this out in his head. "One cent, two cents, four cents, eight cents, sixteen cents, thirty-two cents— geez, after a full week it's still under a dollar a day," he thought. "Okay," he said, eager to tell his father about all the money he had saved himself today. He and the teacher shook hands to close the deal.

Of course, Donald actually lost money with this deal—quite a bit of money. In fact, the payment for the thirtieth day alone was over

$5,300,000 which was 5 million dollars more than the first deal would have cost him—if he hadn't been in such a hurry to dispense with the boring math.

Myth Two
Math is difficult.

Someone with chalk in hand scribbling incomprehensible formulas on a chalkboard can certainly make math seem difficult, but it's not. Like a stepladder, math is composed of steps, each based on a simple idea. You begin on the first step, with arithmetic, and, having mastered the first step, move up to the next. Each step is simple by itself because all the steps depend on the basic building blocks of arithmetic.

Myth Three
Math is too precise.

By itself, precision is nothing to be worried about. Math can be as precise or as imprecise as you need to make it. By becoming comfortable with the basics of math, you can learn to make use of both precision and imprecision. Once you have the fundamentals under your belt, so to speak, you'll find that rounding off, estimating, or educated guesswork is often all that you need to get the job done.

Myth Four
Some people are math types; I'm not.

Most people come to think of themselves as one "type" or another because someone applied the label to them. Of course, some people are naturally better than others at doing certain things, but this never means that they are the *only* people who can do these things or that these are the only things they can do well. Most of us fall into the very large class of people who, collectively, represent the "average" man and woman. We have the native talent, and we can become proficient in math if we wish to pursue it. Our brains nearly always show their enormous power and flexibility if they accept challenges.

Myth Five
Women aren't mathematically inclined.

This is utter nonsense. Of course, differences exist between females and males. Men have greater body strength; women have more endurance. Women can give birth; men cannot. However, men do not have greater natural talent for mathematics than women do. In fact, according to tests given to first graders, girls tend to score slightly higher on math competency than boys do. This probably reflects the fact that girls develop a bit faster than boys in the early years. Yet math competency tests given to graduating high school seniors show a substantial lead for boys.

If boys and girls begin as near equals, why do boys develop better math skills than girls do?[2]

Teachers unconsciously direct greater attention and interest toward boys during math lessons. Teachers of young children are biased, believing that boys are (or should be) more interested in math than girls are. Is this nature or nurture? In this case, we know it's nurture.[3]

Myth Six
Math is not relevant to my life.

Basic mathematical calculations are the oil that keeps the machinery of society running smoothly. If you can't quickly and confidently carry out basic calculations, you are always at the mercy of others. Math is relevant to almost all your daily activities: working, earning money, buying groceries, driving a car, making home repairs, educating children.

Conquering math phobia means gaining power. Millions of Americans embrace the paraphernalia of "ancient and secret"

[2]See Nancy Rubin, "Math Stinks!," *Parents*, June 1988, p. 132; Ursula Casanova and David Berliner, "Are You Helping Boys Outperform Girls in Math?" *Instructor*, October 1987, p. 10; Constance Holden, "Female Math Anxiety on the Wane," *Science*, May 8, 1987, p. 660.

[3]Dr. Myra Sadker and Dr. David Sadker, "Sexism in the Schoolroom of the '80s," *Psychology Today* Vol 19 (March 1985) 54–57.

pseudosciences—crystals, pyramids, tarot cards, and astrology charts—hoping to find increased power and control over their lives. But consider mathematics. Though ancient and, in a sense, secret (because most Americans know so little about it), math *can be mastered* by the average person. More importantly, math is more likely to improve your chances of earning money and of increasing your self-confidence and happiness.

This book *cannot* suddenly erase the negative emotional response you have whenever you need to use math; conquering math phobia may take a little time. What is important now is to recognize and accept your emotional reactions: You are upset by anticipating a bad experience. To help you feel comfortable, this book has very little computation during the introductory chapters.

Your program for achieving math competency is as simple as thinking of walking up a flight of stairs. Read the chapters, and try the exercises. Self-tests at various points in the chapters, with immediately available answers, are provided to guide you in determining your progress. Whenever you feel comfortable, move on. Whenever you are unsure, go back and review. This is *your* program, so move at your own pace; there are no grades—there is only success.

CHAPTER 2

NUMBERS

Wait! Don't slam the book shut! Yes, this chapter is about numbers, but you will *not*, repeat, *not* have to do any computations in this chapter. Instead, you will get some fundamental (and maybe even interesting) information that you will need in order to learn basic operations.

NUMBERS: A BRIEF HISTORY

The first numbers discovered and used by humans were the counting numbers: 1, 2, 3, 4, and so on. Even before the development of farming, primitive hunter-gatherers relied on the use of numbers for their survival. Knowing how many days' travel would lead to the next watering hole was vital information. Knowing how many days a given amount of food would last was important.

The concept of numbers is very old. Carl B. Boyer suggests in his book, *A History of Mathematics,* that the discovery of counting and numbers probably goes back as far as mankind's use of fire, roughly 300,000 years.[1] Some early cultures had names for only three distinct whole numbers: 1, 2, and 3. Larger collections were called "many" and were counted in groupings of 3.

The Babylonians and Egyptians were the first to develop comprehensive mathematical systems in the second millenium B.C., but it was left to the Greeks to establish a theoretical foundation for mathematics. The early Greeks did not use written symbols for numbers;

[1] Carl B. Boyer, *A History of Mathematics, Second Edition,* (New York: John Wiley and Sons, 1968, 1990) 3.

they used what was called pebble notation (see Figure 1). To write the number 1, you simply put down a dot or pebble. To write 2, you put two pebbles side by side. Because the number 3 can be formed by putting three dots into the shape of a pyramid, the early Greeks called 3 a pyramid number (sometimes referred to as a triangle number). Similarly, the number 4 became a square number.

The Greeks discovered, as their need to record calculations increased, that it was more convenient to assign a unique symbol to each number than to write down masses of dots. However, pebble notation can still be helpful, as you will see in later computation chapters.

In fact, the Greek number notation was awkward and difficult to use in calculations, and the numerals introduced by the Romans were even more awkward. Because of this, the *abacus*, an ancient mechanical device for computing, was in common use in most parts of the world for many centuries during and following the height of Greek and Roman civilizations.

Our modern number system originated in India and was first introduced into Europe through the Arabs in the tenth century. This system, called the Hindu-Arabic system, gained wide acceptance in the thirteenth century and fully replaced Roman numerals in mathematics in the sixteenth century.

The Hindu-Arabic numbering system uses 10 unique symbols, called digits, to write the numbers 0 through 9. To move up from 9, a 1 is placed to the left of the single digit and a 0 replaces the 9

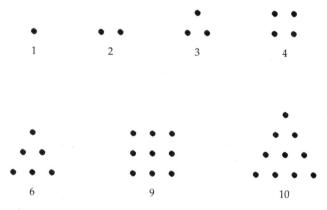

Figure 1.   Ancient Greek pebble, or dot, notation.

to form *10*. Here, the zero is considered a placeholder and prevents confusion of the number 10 with the number 1. To write the number after 10, change the 0 to 1; you now have 11. Write the next number by changing the right digit 1 to a 2; you have 12. The position that 2 occupies is called the *ones place*, and the position that 1 occupies is called the *tens place*. When you reach 19, add another 1 to the 1 already in the tens place; write 2 in the tens place and 0 in the ones place to form 20. Performing basic calculations with this system is much easier than with the Roman system.

Consider the number 792. The digit 2 occupies the ones place, 9 occupies the tens place, and 7 occupies the hundreds place. You can break this number down into sums that make the number easier to understand: 792 = 700 + 90 + 2. If you write a number with four digits, such as 7,920, the leftmost digit (7) occupies the thousands place. This process can go on indefinitely as digits occupy the ten thousands place, the hundred thousands place, and so on.

PROBLEMS

Identify the position or positions (ones place, tens place, hundreds place, or thousands place) occupied by the digit 1 in the following numbers.

1. 17

2. 301

3. 4,102

4. 1,012

5. 71,441

ANSWERS

1. In 17 the 1 occupies the tens place.

2. In 301 the 1 occupies the ones place.

3. In 4,102 the 1 occupies the hundreds place.

4. In 1,012 the 1 occupies both the thousands place and the tens place.

5. In 71,441 the 1 occupies both the thousands place and the ones place.

NEGATIVE NUMBERS, OR LOANS FROM THE BANK

The numbers discussed so far have been positive numbers. Negative numbers also exist. To approach negative numbers, look at Figure 2, which shows part of a line that stretches forever in two directions. Let's choose a point on this line to represent the number 0. Next, move a fixed distance to the right of 0 and put in a point to represent the number 1. Figure 2 shows a number line with zero and the positive numbers 1 through 5. Negative numbers will be on the number line to the left of 0. The plus signs in front of each positive number and the minus signs in front of each negative number remind you that positive numbers are different from negative numbers.

Why do you need to know about negative numbers? You have an intuitive sense for what positive numbers are because we use them to count, but what does negative five mean? Think of numbers as money (since that's what money really is): negative numbers would represent loans from the bank (money you owe), whereas positive numbers would represent money you have deposited into the bank. If you have five dollars in your savings account, you have +5 dollars (and probably a letter from the bank suggesting a cash infusion), but if you owe the bank five dollars, you have −5 dollars (and probably a more strongly worded letter from the bank).

To avoid confusion when adding or subtracting, we include a negative sign when writing a negative number but use no sign when writing a positive number. Since the negative sign could be easily confused with the sign that indicates subtraction, we use parentheses to enclose negative numbers:

$$5 + (-3) = 2$$

An equal sign (=) shows that the numbers or expressions on both sides of the sign have the same value; "5 + (−3)" has exactly the same value as "2" (but one takes longer to say).

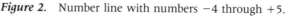

Figure 2. Number line with numbers −4 through +5.

WHAT HAVE YOU LEARNED?

Before you move on to the next chapter, you should know the following:

1. Our current numbering system, called the Hindu-Arabic system, uses the ten symbols 1, 2, 3, 4, 5, 6, 7, 8, 9, and 0. Numbers are formed by using the digits in combination. Each position has a name beginning with the rightmost position, called the ones place. The next three positions are called the tens place, the hundreds place, and the thousands place.
2. The positive numbers can be represented by points on a line to the right of 0.
3. The negative numbers can be represented by points on a line to the left of 0.
4. The plus sign (+) and the minus sign (−) can be used to indicate whether a number is positive or negative or to indicate the operation of addition or subtraction. Use parentheses to avoid confusing the operation signs with the signs indicating negative numbers.

You have accomplished much, for you now know about positive and negative whole numbers, the zero, the plus (+) and minus (−) signs, and the equal sign (=), the last three of which are modern developments in mathematical notation. Understanding this much will help you through the next chapter.

CHAPTER 3

MORE NUMBERS

In this chapter, you will examine two other kinds of numbers, fractions and decimals—probably the most dreaded of all numbers. But when you have finished this chapter, not only will you have reviewed all the kinds of numbers you'll need to know to do arithmetic, but you'll have done so painlessly. (If you're already familiar with fractions and decimals, you might want to try the problems at the end of each section of this chapter; if you understand these and answer them all correctly, you can go to Chapter 4.)

FRACTIONS

First, consider fractions. Anything but fractions, right? Fractions *do* tend to frighten people. Gruesome though they may seem at first, however, fractions are a fundamental part of our number system and serve many useful purposes. You can manage fractions if you proceed slowly and carefully.

You use fractions frequently in recipes. If your recipe is for twelve people, but five of your anticipated dinner guests cancel at the last minute, how do you change the amounts given by the recipe? (And how can you get revenge on those five so-called friends? But that's another chapter.) Suppose the original recipe called for 1 ¾ cups of water. How would you adjust that for only seven guests? What you need to know for such problems is how to add, subtract, multiply, and divide fractions.

Fractions are written with two whole numbers, one number above the other, with a horizontal or diagonal line dividing them.

$$\frac{\text{Top whole number}}{\text{Bottom whole number}} \quad \text{or} \quad \frac{3}{4} \quad \text{or} \quad {}^{3}\!/_{4}$$

In printed texts, fractions are frequently written straight across on one line: the top number is first, followed by a slash, followed in turn by the bottom number.

Top whole number/Bottom whole number   or $^{3}\!/_{4}$

Each of these two whole numbers tells you something about the fraction they represent. The bottom number tells you how many equal parts to think about. The top number tells you how many of these equal parts the fraction has. Confusing? You bet. But it's actually easy. Suppose you have baked an apple pie and want to share it with two friends. You want to separate the pie into three equal parts. Each of you will receive an equal share, or one-third, of the pie. The fraction that represents the amount each will receive is $^{1}\!/_{3}$. The number 3 in this fraction shows that you divided one whole pie into 3 equal parts.

Fractions can also be understood in relation to the number line we discussed in Chapter 2. You can see at once from Figure 2 that we haven't used all the positions on the line; there are equal-sized line segments between whole numbers. Beginning at 0, move slowly right toward the point representing the number 1. When you get halfway, stop. You're halfway between the numbers 0 and 1. Make a mark, and call it the number $^{1}\!/_{2}$. Life experience tells you what $^{1}\!/_{2}$ is: half of an apple pie, half of the money in Fort Knox, or half of anything you want.

It's important not to confuse the number $^{1}\!/_{2}$ with the two numbers 1 and 2; $^{1}\!/_{2}$ is a number all by itself.

If you continue moving toward the right from $^{1}\!/_{2}$, you can stop halfway between $^{1}\!/_{2}$ and 1. Put a point there, and call this point the fraction $^{3}\!/_{4}$. Imagine the distance between 0 and 1 to be divided into four equal parts. You have moved across three of them. The 3 and the 4 in this fraction tell you that you have covered 3 of 4 equal parts.

You can, in fact, form any fraction that has a positive number above and a larger positive number below, and that fraction will always fall between 0 and 1. For example, you could locate the fraction $^{1,021}\!/_{76,632}$

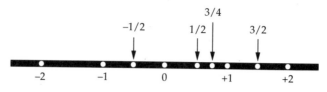

Figure 3.   Number line containing fractions.

by dividing the distance between 0 and 1 into 76,632 equal parts and then moving across the first 1,021 of them.

Now move to the right on the number line until you are halfway between 1 and 2. Make a point there, and call it one and one-half, or 1 ½. What is the fraction for 1 ½? It is the 2 equal lengths between 0 and 1 plus 1 more of those lengths. Because the distance between 0 and 1 is divided into two equal lengths, the fraction's bottom number is 2. Because there are three of these lengths—2 between 0 and 1 plus 1 more—the top number is 3. The fraction is ³⁄₂. Simple, isn't it?

In the fraction ³⁄₂ the top number is greater than the bottom number. In fact, you can write a fraction with any two whole numbers, positive or negative. It doesn't matter which is greater. The bottom number tells how many equal lengths are between 0 and 1. The top number tells the number of these lengths that, placed end to end, will locate the fraction. For example, you can locate the fraction ⁶⁴⁄₇ in the following way: Divide the distance between 0 and 1 into 7 equal lengths, then connect a total of 64 of these segments end to end toward the right and see how far they reach. The fraction is located between the numbers 9 and 10.

For every positive fraction you place on the line, you can also place a negative fraction. Negative fractions are located to the left of 0 rather than to the right.

So far, you have constructed fractions with whole numbers. But you cannot construct fractions with the number 0. If you encounter a fraction whose top number is 0, then it doesn't matter what the bottom number is; that fraction is equal to 0. ⁰⁄₅ = ⁰⁄₁,₅₅₉,₈₆₇ = ⁰⁄₆₅ = 0.

Rule
A fraction whose top number is 0 is equal to 0.

What if the bottom number is 0? It's impossible. If in your calculations you see a fraction with 0 as the bottom number, you know some calculation was done improperly, and the calculations should be repeated to eliminate the error.

Rule
There exists no fraction that has a bottom number of 0.

You now have a definition for a fraction.

Definition
A fraction is a number written as a whole number divided by a second whole number; neither number is 0.

You can also write any whole number as a fraction. We do this by writing the whole number as the top number and the number 1 as the bottom number. You can write 2 as $^2/_1$—or 17,442 as $^{17,442}/_1$. You usually won't write them this way because it simply takes too much time. Remember that mathematicians are notoriously lazy. If they can find a shorthand way to do something, they'll do it. That's good for you because it makes your calculations easier if you take the time to learn their shorthand.

Fractions can be either positive or negative. Because fractions are made of two numbers and each number can be either positive or negative, four possible combinations exist. For example, using a 1 and a 2 to construct fractions, you can have the following combinations.

$$1/2 \qquad (-1)/2 \qquad 1/(-2) \qquad (-1)/(-2)$$

Note the parentheses around negative numbers to avoid confusion. You already know that the first fraction, 1/2, is a positive fraction and belongs to the right of zero, halfway between the numbers 0 and 1. What about the other three? The second and third fractions, $(-1)/2$ and $1/(-2)$, are both negative fractions because they include a single negative number, and they therefore belong to the left of zero.

$$(-1)/2 = -(1/2) \qquad \text{and} \qquad 1/(-2) = -(1/2)$$

$$\text{therefore, } (-1)/2 = 1/(-2)$$

The last fraction, $(-1)/(-2)$, is the positive fraction 1/2.

$$(-1)/(-2) = 1/2$$

Here are the basic rules for determining the sign of a fraction.

Rule

If the top number and the bottom number of a fraction have the same sign (both positive or both negative), the resulting fraction is positive.

Rule

If the top number and the bottom number of a fraction have opposite signs (one negative and the other positive), the resulting fraction is negative.

PROBLEMS

Identify which of the following fractions are positive fractions and which are negative fractions.

1. (-1)/3
2. (-2)/(+1)
3. 41/36
4. (-155)/(-621)
5. 7/(-13)

ANSWERS

1. (-1)/3 is negative.

2. (-2)/(+1) is negative.

3. 41/36 is positive.

4. (-155)/(-621) is positive.

5. 7/(-13) is negative.

DECIMALS

A decimal is not another type of number; it is merely another way of writing a fraction or a whole number. Like whole numbers and fractions, decimals can be positive or negative. Here are some examples of decimals with their corresponding fractions or whole numbers.

¼	.250
7	7.0
⅔	2.0000
½	0.5
⅓	.33333...

You probably noticed that one of the decimals, the one representing ⅓, ends at the right with three periods. Don't let this throw you. There are two kinds of decimals. The simpler, and by far more common, is the terminating decimal, of which the first four decimals shown are examples.

Definition
A terminating decimal is a decimal that contains a finite number of digits.

Consider the decimal 0.5, which stands for the fraction $^5/_{10}$, or ½. In Chapter 9 you will learn to easily change one fraction into another. You can write this decimal in a variety of ways.

$$.5 = 0.5 = 00.5 = 0.50 = 0.5000 = .50000000$$

Each of these forms stands for the fraction ½. When you write zeros to the left of the decimal point, you don't change the value of the number. Nor do you change the value by writing more zeros at the right. What you must keep unchanged are the nonzero digits, any zeros between these digits, and the decimal point. Although you will see decimals less than 1 written with and without zero to the left of the decimal point (for example, .5 and 0.5), you may want to make a rule for yourself to write that zero. It helps you to remember that the number is less than a whole number.

Definition
An infinite decimal is a decimal that contains an infinite number of digits.

Many infinite decimals have repeating digits. The fraction ⅓ is represented by the infinite, repeating decimal 0.3333..., where the three periods at the right indicate that the threes go on forever. In this case, the repeating-digit sequence consists of only one digit: 3. Of course, you could never write such a decimal with *all* the threes, but you know what the decimal is and you know that it is equal to ⅓. Some repeating decimals repeat a longer sequence of digits. Here is an example:

$$0.142857142857142857\ldots$$

This decimal, which stands for the fraction ⅐, repeats the sequence 142857 indefinitely.

Rule
Every fraction and whole number can be represented by either a terminating decimal or an infinite, repeating decimal.

Rule
Every terminating decimal and infinite, repeating decimal can be represented by either a fraction or a whole number.

The fractions ½ and ⅕ are represented by terminating decimals (0.5 and 0.2), and the fractions ⅓ and ⅐ are represented by infinite, repeating decimals (0.333... and 0.142857142857...).

What will you do with an infinite decimal when you need to use it in a calculation? You will have to chop it off somewhere to the right of the decimal point and treat it as a terminating decimal. Where you chop it off (or round it) depends on how precise you want to be. In Chapter 16 you will be shown how to round off numbers to increase the accuracy of your calculations.

Rule
When calculating with infinite decimals, round off the decimal to use it as a terminating decimal.

Decimals are an extension of the place-value system you use to write whole numbers. In a whole number, each digit occupies a specific place that tells you the value of that digit. For example, a 1 standing alone is 1, but a 1 at the left of a zero (10) is occupying the tens place, so you know that the number is 10. Each time you move a place toward the left (10, 100, 1,000, and so on), you increase the value of the symbol 10 times because our numbering system is based on 10. In like manner each time you move one place toward the *right* of the decimal point (.1, .01, .001, and so on), you decrease the value of the symbol to $\frac{1}{10}$. Hence, .01 has a value $\frac{1}{10}$ the value of .1.

Suppose you have sent your son to the store to pick up 4 quarts of milk, but he comes home with only 3 ½ quarts. How can you find out what happened to the other ½ quart? More to the point, how can you write this number as a decimal? Write 3 in the ones position, representing the 3 complete quarts. Write the part representing less than a full quart to the right of the 3, which is the tenths place. Because ½ expressed in tenths is $\frac{5}{10}$, the digit you write at the right of the decimal point is 5. So 3 ½ quarts is 3.5 quarts. Using *expanded notation,* you can write a decimal as a sum of the digits occupying the various positions.

$$437.194 = 400 + 30 + 7 + .1 + .09 + .004$$

With a full understanding of the values of the places, much that seems confusing about math will become clearer, and you can avoid getting hopelessly muddled when you try to manipulate numbers. At this point, don't worry about manipulating decimals or changing a fraction into a decimal. This chapter is meant to introduce you only to the *idea* of decimals. The procedures for manipulating them will be covered in easy steps in Chapters 9 through 12.

PROBLEMS

Identify each number below as a positive whole number, a negative whole number, 0, a positive fraction, a negative fraction, a positive decimal, or a negative decimal.

1. (-2)

2. 33

3. $\frac{3}{9}$

4. .5

5. 1,772

6. (-2.55)

7. $\frac{9}{7}$

8. -($\frac{3}{4}$)

9. 0.00

10. (-2.4777 ...)

ANSWERS

1. The number (-2) is a negative whole number.

2. The number 33 is a positive whole number.

3. The number ⅜ is a positive fraction.

4. The number .5 is a positive decimal.

5. The number 1,772 is a positive whole number.

6. The number (-2.55) is a negative decimal.

7. The number ⁰⁄₇ is 0.

8. The number -(¾) is a negative fraction.

9. The number 0.00 is 0.

10. The number (-2.4777...) is a negative decimal. It is also an infinite, repeating decimal—notice the three periods.

If you missed more than three of the problems, review the chapter and try again.

WHAT HAVE YOU LEARNED?

You have now met every member of the number family you will ever need for arithmetic: positive and negative whole numbers, and their more complicated offspring, fractions and decimals. You haven't had to manipulate them yet. Therefore, it's understandable if you still feel a little queasy about fractions and decimals. Don't worry; the methods used to manipulate numbers will be clearly explained in the following chapters.

Here is a recap of some rules and definitions you will need for arithmetic computations:

1. A fraction whose top number is zero (⁰⁄₄) is equal to zero.
2. There exists no fraction that has a bottom number of zero (⁴⁄₀ = problem; try again).
3. A fraction is a number written as a whole number divided by a second whole number (⅓ = 1 ÷ 3); neither number is zero.
4. If the top number and the bottom number of a fraction have the same sign (¾ [both positive] or -3/-4 [both negative]), the resulting fraction is positive.

5. If the top number and the bottom number of a fraction have opposite signs (-3/4 or 3/-4), the resulting fraction is negative.
6. Decimals are an alternative way of writing fractions and whole numbers (5 ½ = 5.5).
7. Two kinds of decimals exist: terminating decimals (0.31) and infinite decimals (0.333...).
8. A terminating decimal has a finite number of digits.
9. An infinite decimal has an infinite number of digits. If a set of digits repeats indefinitely, then the infinite decimal is an infinite, repeating decimal. An infinite, repeating decimal is written as a terminating decimal followed by three dots or periods (the infinite, repeating decimal for the fraction ⅓ is written as 0.3333...).
10. Every fraction or whole number can be written as either a terminating decimal (¼ = 0.25; 4 = 4.0) or an infinite, repeating decimal (⅐ = 0.142857142857142857...).
11. Every terminating decimal or infinite, repeating decimal can be written as a fraction or a whole number.
12. When calculating with an infinite decimal, round off the decimal to use it as a terminating decimal (0.142857142857142857... can be rounded to 0.1429, 0.143, or 0.14). For rounding off numbers, see Chapter 16.

TAKING STOCK

Take a moment to reflect on what you have learned. Right now, believe it or not, you probably know more about numbers than the vast majority of people who have lived before you. When you have finished this book, you will know more about mathematics than the vast majority of people alive today.

CHAPTER 4

MASTERING ADDITION

Now that you have the different kinds of numbers under your belt, you can begin learning how to use them comfortably. If you consider your addition skills strong already, you may want to skim this chapter and try your hand at the addition problems at the end of each section.

SOME WAYS TO THINK OF ADDITION

Addition is not only the most common mathematical operation, but also the most important. In a sense, addition defines the positive numbers. Each positive number is defined as the sum of the number 1 and the preceding number. Three can be defined as 2 plus 1, four as 3 plus 1, and so on.

Adding is like combining groups of things. If your nephew has a collection of five marbles, and you give him two more, he has a collection of seven marbles. (Remember pebble numbers from Chapter 2?) Soon, he'll forget all about marbles and begin to use only numbers, as you're doing now.

You can also think of addition in terms of the straight line used in Chapters 2 and 3. You add numbers by combining lengths of line. If you wish to add 3 plus 2, begin at 0, and move to the right 3 units; then move two more.

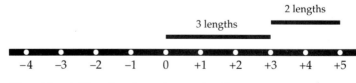

Figure 4. Adding 2 to 3 on the number line.

SHORT ADDITION

In this section, you'll be adding only the numbers 0 through 10. You have two terrific shortcuts: counting on your fingers and memorizing the answers.

Many adults are embarrassed by using their fingers to count. They hide their hands under tables and try not to move their lips as they count. But counting on your fingers is perfectly acceptable. Each morning, I cook a six-minute egg. After dropping the egg into boiling water, I add six minutes to the time shown on the kitchen clock. I don't feel comfortable unless I count off six fingers against the clock. If you feel comfortable counting on your fingers, go ahead—and know that you're definitely not alone.

The second shortcut is to memorize the answers. To work easily with addition, you should memorize the sums of whole numbers less than 10 until you can rattle them off without thinking. Although counting on your fingers is a perfectly acceptable method, it is just too slow.

Developing ease and comfort with the addition of numbers below 10 is important because most computing is done with those numbers. Table 4.1 is an addition table for the numbers below 10. To add a number shown at the left (5, for example) to a number shown at the top (say, 7), look for the result in the intersection of the row and column (in this case, 12).

Half of the available spaces in Table 4.1 are blank. Because $7 + 5 = 5 + 7 = 12$, you need to show the result (12) only once. Therefore, you

Table 4.1 Addition Table for Numbers under Ten

+	2	3	4	5	6	7	8	9
2	4	5	6	7	8	9	10	11
3		6	7	8	9	10	11	12
4			8	9	10	11	12	13
5				10	11	12	13	14
6					12	13	14	15
7						14	15	16
8							16	17
9								18

can cut the table in half simply by remembering that $A + B$ is the same as $B + A$. This characteristic of addition, known as the commutative property, simply means that the order in which you add two numbers does not matter. (If the word *commutative* seems too technical, think of it as the cash-register property: At the checkout line, it doesn't matter if the salesperson rings up your goods as pepperoni, onions, and beer, or onions, beer, and pepperoni, or beer, onions, and pepperoni; the total will be the same, and your breath will be equally lethal, regardless.)

Since the number 1 added to any number is equal to the next number in the number sequence, the table does not include the 1 column or row. Nor does the table include 10, because 10 plus any number less than 10 is written as a 1 followed by the number added, such as: $10 + 5 = 15$. This makes addition with 10 so easy that you don't really have to memorize it. Thus, you have a table with only 36 entries to memorize. Zero is not included, because adding zero to a number is entirely futile: No matter how many times you try or how experienced you are at it, adding zero will never change a number's value.

You have probably memorized most of these 36 sums already. By memorizing all of them, your progress up the mathematics stairway will be greatly eased. How do you do this? Repetition. For help, you can buy flash cards sold in school supply stores. You can also practice by playing card games: Blackjack, or 21, is an excellent game for practicing addition.

PROBLEMS

Without looking at Table 4.1 or using your fingers, add each pair of numbers.

1. $4 + 3 = ?$	**6.** $6 + 9 = ?$	**11.** $7 + 8 = ?$
2. $3 + 8 = ?$	**7.** $5 + 5 = ?$	**12.** $0 + 3 = ?$
3. $10 + 1 = ?$	**8.** $2 + 6 = ?$	**13.** $3 + 9 = ?$
4. $1 + 6 = ?$	**9.** $7 + 2 = ?$	**14.** $5 + 4 = ?$
5. $0 + 9 = ?$	**10.** $0 + 10 = ?$	**15.** $6 + 7 = ?$

ANSWERS

1. 7	**6.** 15	**11.** 15
2. 11	**7.** 10	**12.** 3
3. 11	**8.** 8	**13.** 12
4. 7	**9.** 9	**14.** 9
5. 9	**10.** 10	**15.** 13

BREAKING NUMBERS DOWN

The numbers 0 through 9 are called digits, and larger numbers are made up of combinations of digits. You can take any number and break it into its parts. For example, you can break the number 12 into 10 + 2.

Suppose you have only $5 in your wallet, but Jim down the street is on his way over to repay that $12 he owes you, and you want to figure out how much spending money that will give you. You want to add 5 to 12, but the addition table (Table 4.1) does not show the answer—what do you do? You can apply the answers in the addition table to separate groups of digits. You can break 12 into two parts (10 + 2) and then add the 5.

$$12 + 5 = (10 + 2) + 5$$

Now group the 2 and the 5 together.

$$12 + 5 = 10 + (2 + 5)$$

Remember that 2 plus 5 is 7. So

$$12 + 5 = 10 + 7$$

You know that 10 combined with 7 is 17. So

$$12 + 5 = 17$$

After memorizing Table 4.1, you can add numbers of almost any size by simply breaking the numbers down into smaller parts, adding the parts from memory, and recombining them.

The next step is to recognize that 10 plus 10 is 20, and that 30 plus 40 is 70. Use the table to calculate that 3 plus 4 is 7, allowing the 0 to hold its place. In the same way, you add 9,000 to 6,000 for 15,000 (9 + 6 = 15). Here the zeros are placeholders, while the 9 and the 6 are combined to form 15.

PROBLEMS

Add the following pairs of numbers in your head.

1. 11 + 3 = ?	**6.** 30 + 40 = ?
2. 14 + 6 = ?	**7.** 100 + 400 = ?
3. 10 + 70 = ?	**8.** 110 + 30 = ?
4. 600 + 200 = ?	**9.** 700 + 900 = ?
5. 50 + 80 = ?	**10.** 15, 000 + 2, 000 = ?

ANSWERS

1. 14	**6.** 70
2. 20	**7.** 500
3. 80	**8.** 140
4. 800	**9.** 1,600
5. 130	**10.** 17,000

"Terrific," you say, "but what do I do with larger, more complicated numbers?" What you need is a method that is both fast and easy— long addition, which allows you to break a large problem into many small problems.

LONG ADDITION

Long addition allows you to add any two positive numbers. You don't have to memorize any more answers from an addition table because a clever little trick will let you avoid all that. It's called *carrying*. Begin with two 2-digit numbers. Let's say you need to quickly figure out how many hours you worked last week: You know you put in a total of 23 hours on Monday, Tuesday, and Wednesday; then you took off Thursday and Friday as Personal Sanity days but worked 14 hours over the weekend.

$$23 + 14 = ?$$

First stack these numbers and draw a line under them.

$$
\begin{array}{r}
2\ 3 \\
+\ 1\ 4 \\
\hline
\end{array}
$$

Begin with the two digits stacked on the right, the 3 and the 4. Adding these from your memory, you know that the answer is 7. Write the 7 under the line, directly beneath the 3 and the 4.

$$
\begin{array}{r}
2\ 3 \\
+\ 1\ 4 \\
\hline
7
\end{array}
$$

Add the 2 and the 1; write the 3 under the line.

$$
\begin{array}{r}
2\ 3 \\
+\ 1\ 4 \\
\hline
3\ 7
\end{array}
$$

Thus, you put in 37 hours last week. This is simple (the math, I mean, not the hours). But what if the sum of the two stacked digits is greater than 9? Here, you must learn the trick of *carrying*. What if you really knocked yourself out last weekend and worked 19 hours?

```
    2 3
  + 1 9
  ─────
```

First add the digits at the right, 3 and 9, to get 12. Your first impulse might be to write 12 below the line, but that would lead you astray. Break 12 into 10 + 2. Now you can proceed. Instead of writing 12 below the line, write only the 2.

```
    2 3
  + 1 9
  ─────
      2
```

What do you do with the 10 left over? Stack it on top.

```
   (1 0)
    2 3
  + 1 9
  ─────
      2
```

The 10 is in parentheses only to remind you that it was carried and not part of the original problem. Now add the left column of digits, remembering to add the 1 digit that was carried. Adding 1 plus 2 plus 1 gives you 4, so write 4 below the line.

```
   (1 0)
    2 3
  + 1 9
  ─────
    4 2
```

The sum of 23 and 19 is 42. You didn't even have to memorize it or count on your fingers (and toes). You can simplify the process even more by not carrying the entire number—in this case, 10. Instead, break 12 into its digits, and carry the 1, or the tens digit, to the top of the stacked numbers. Here is the simplified long addition.

```
    (1)
    2 3
  + 1 9
  ─────
    4 2
```

The parentheses aren't necessary; think of them as a reminder of the process.

Rule

(Carrying) In long addition, if a column adds up to 10 or more, write the ones digit under the line, and add the tens digit to the column on the left.

Now let's say you and your spouse go out for a special dinner to celebrate the promotion you got for working so hard. The total for food is $67, and the total from the bar (a bottle of champagne) is $46. The check says you owe $123 before tax. Is it right?

$$\begin{array}{r} 6\ 7 \\ +\ 4\ 6 \\ \hline \end{array}$$

First, add the 7 and the 6 to get 13. Split the 13 into 10 and 3. Place the 3 below the line, and carry the 10 (but write only the digit 1 and ignore the 0).

$$\begin{array}{r} (1) \\ 6\ 7 \\ +\ 4\ 6 \\ \hline 3 \end{array}$$

Now add the 1 that was carried to the 6 and the 4. The sum of 6 and 4 is 10; adding the carried 1 makes the sum 11.

$$\begin{array}{r} (1) \\ 6\ 7 \\ +\ \ \ 4\ 6 \\ \hline 1\ 1\ 3 \end{array}$$

The sum of 67 and 46 is 113. Call this to the waiter's attention, and your new math skills have saved you $10 already.

At this point, you can practice long addition of numbers less than 100.

PROBLEMS

Add the following pairs of numbers.

1. 13 + 11 = ?

2. 35 + 41 = ?

3. 97 + 16 = ?

4. 43 + 20 = ?

5. 62 + 19 = ?

6. 19 + 7 = ?

7. 52 + 54 = ?

8. 81 + 21 = ?

9. 99 + 99 = ?

10. 44 + 71 = ?

ANSWERS

1.   1 3
    + 1 1
    ─────
      2 4

2.   3 5
    + 4 1
    ─────
      7 6

3.   (1)
        9 7
    +   1 6
    ───────
      1 1 3

4.   4 3
    + 2 0
    ─────
      6 3

5.   (1)
      6 2
    + 1 9
    ─────
      8 1

6.   (1)
        1 9
    +   7
    ───────
        2 6

7.   5 2
    +   5 4
    ───────
      1 0 6

8.   8 1
    +   2 1
    ───────
      1 0 2

9.   (1)
        9 9
    +   9 9
    ───────
      1 9 8

10.   4 4
    +   7 1
    ───────
      1 1 5

For additional practice, you can simply make up your own problems and then check your answers with a calculator. You might say, "Aha! Why can't I just do it on a calculator? Why do I have to learn the procedure?" The answer is simple: Calculators are a convenience, but they are fallible because you can sometimes enter numbers incorrectly. More important, most of the times when you'll need addition (as in the restaurant or at the checkout line), you probably won't have access to a calculator.

Now you can move on to bigger numbers. Let's say Billie Jean, a *very* ambitious teen-ager saving up for a trip to a Michael Jackson concert, earns $147 per week at her summer day job and decides to take on part-time evening work at the local fast-food restaurant. Her evening work pays $78 per week. How much will Billie Jean earn per week?

$$
\begin{array}{r}
1\ 4\ 7 \\
+\ \ \ 7\ 8 \\
\hline
\end{array}
$$

Add the two digits stacked at the right: 7 added to 8 is 15; 15 can be broken into 10 and 5. Place the 5 below the line and carry the 10.

$$
\begin{array}{r}
(1) \\
1\ 4\ 7 \\
+\ \ \ 7\ 8 \\
\hline
5
\end{array}
$$

Next add the 4 to the 7, plus the carried 1; 4 plus 7 is 11, and 1 more is 12. Break the 12 into 10 and 2. Write the 2 below the line in the tens place and carry the hundreds digit (1).

$$
\begin{array}{r}
(1)(1) \\
1\ 4\ 7 \\
+\ \ \ 7\ 8 \\
\hline
2\ 5
\end{array}
$$

Now add the column at the far left, 1 plus 1. Write the result, 2, below the line.

$$
\begin{array}{r}
(1)(1) \\
1\ 4\ 7 \\
+\ \ \ 7\ 8 \\
\hline
2\ 2\ 5
\end{array}
$$

Billie Jean will be earning $225 per week if she works both jobs.

The next example is a little harder. Let's say your town, with a population of 76,983 residents, is going to annex a neighboring community with 91,701 people. What will the combined population be?

$$\begin{array}{r} 7\,6,9\,8\,3 \\ +\,9\,1,7\,0\,1 \\ \hline \end{array}$$

Add the far right column, 3 plus 1, to get 4. Because 4 is less than 10, you don't have to break it apart or carry anything. You can simply write 4 below the line.

$$\begin{array}{r} 7\,6,9\,8\,3 \\ +\,9\,1,7\,0\,1 \\ \hline 4 \end{array}$$

Add the next column to the left; add 8 to 0 to get 8. Since this number is also less than 10, simply write it below the line.

$$\begin{array}{r} 7\,6,9\,8\,3 \\ +\,9\,1,7\,0\,1 \\ \hline 8\,4 \end{array}$$

Now add the next column, 9 and 7, to get 16. Because 16 is greater than 9, break it into 10 and 6, write the 6, and carry the 10.

$$\begin{array}{r} (1) \\ 7\,6,9\,8\,3 \\ +\,9\,1,7\,0\,1 \\ \hline 6\,8\,4 \end{array}$$

Add the next column, 6 plus 1, plus the carried 1, which equals 8. You have nothing to carry. Write the 8.

$$\begin{array}{r} (1) \\ 7\,6,9\,8\,3 \\ +\,9\,1,7\,0\,1 \\ \hline 8\,6\,8\,4 \end{array}$$

Add the last column, 7 and 9, to get 16. Because you have no more columns at the left to add, you can write 16 without breaking it apart.

$$
\begin{array}{r}
(1)\\
7\,6,9\;8\;3\\
+\,9\,1,7\;0\;1\\
\hline
1\;6\,8,6\;8\;4
\end{array}
$$

Try the following exercises. Sample Problems: Add each pair of numbers.

PROBLEMS

Add each pair of numbers.

1. $\begin{array}{r} 1\;8\;3 \\ +\,4\;2\;7 \\ \hline \end{array}$

2. $\begin{array}{r} 7\;3\;6 \\ +\,6\;1\;0 \\ \hline \end{array}$

3. $\begin{array}{r} 9\;0\;2 \\ +\,9\;9\;1 \\ \hline \end{array}$

4. $\begin{array}{r} 1\;1\;4 \\ +\;\;\;7\;7 \\ \hline \end{array}$

5. $\begin{array}{r} 8\;9\;4 \\ +\,1\;0\;9 \\ \hline \end{array}$

6. $\begin{array}{r} 8\;9\;0 \\ +\,4\;3\;2 \\ \hline \end{array}$

ANSWERS

1. (1)(1)
$\begin{array}{r} 1\;8\;3 \\ +\,4\;2\;7 \\ \hline 6\;1\;0 \end{array}$

2. $\begin{array}{r} 7\;3\;6 \\ +\;\;6\;1\;0 \\ \hline 1,3\;4\;6 \end{array}$

3. $\begin{array}{r} 9\;0\;2 \\ +\;\;9\;9\;1 \\ \hline 1,8\;9\;3 \end{array}$

4. (1)
$\begin{array}{r} 1\;1\;4 \\ +\;\;\;7\;7 \\ \hline 1\;9\;1 \end{array}$

5. (1)(1)
$\begin{array}{r} 8\;9\;4 \\ +\;\;1\;0\;9 \\ \hline 1,0\;0\;3 \end{array}$

6. (1)
$\begin{array}{r} 8\;9\;0 \\ +\;\;4\;3\;2 \\ \hline 1,3\;2\;2 \end{array}$

ADDITION BY COLUMNS

One of the most useful techniques in arithmetic is adding columns of numbers—long addition that involves adding more than two digits at a time. Adding by columns is frequently used to check the work of others, especially when checking bills and receipts. Reviewing your bills through column addition can save you money at once.

Begin with a simple problem. In three different trips to the grocery store last week, you spent $27, $14, and $54. What was the total amount you spent on groceries last week?

$$\begin{array}{r} 2\ 7 \\ 1\ 4 \\ +\ 5\ 4 \\ \hline \end{array}$$

Always begin with the right-hand column. Work with only two digits at a time. Begin by adding 7 + 4 to get 11. Next, add the 4 to the 11 to get 15. "But wait," you say. "The sum of 11 and 4 is not in the addition table. That's another long addition problem." You're quite right. But by now, you know that adding any single-digit number to any other number is like counting. So even in column addition, you never have to add more than two numbers at a time. Use a "scratch pad" if you find it helpful.

(Main Problem)   (Scratch Pad)

$$
\begin{array}{rl}
(1) & \\
27 & \\
14 & \qquad 11 \quad \text{the sum of 7 and 4 in the right column} \\
+\ 54 & \qquad +\ \ 4 \\
\hline
95 & \qquad \overline{15}
\end{array}
$$

Thus, you spent $95 on groceries last week. Now you're ready to try a tougher problem (or go on a diet). To cut back on weight and the grocery bill, you begin to count calories. Your three meals for your first dieting day contained 892, 431, and 706 calories, respectively. What was your calorie count for the day? Get out your calorie counter and your scratch pad.

(Main Problem)	(Scratch Pad)
(1)	12 the sum of 8 and 4 in left column
8 9 2	+ 7
4 3 1	19
+ 7 0 6	+ 1 add the 1 carried to the left column
2, 0 2 9	20

On your first day of dieting, you took in 2,029 calories. Note that the scratch pad shows the addition of the far-left column. You may find it helpful to begin the column addition with the carried number. For the left column we have $1 + 8 = 9, 9 + 4 = 13$, and $13 + 7 = 20$.

Try one more example, without using the scratch pad.

$$4, 2\ 9\ 0$$
$$6, 6\ 1\ 9$$
$$7, 4\ 0\ 4$$
$$+\ 3, 6\ 5\ 2$$

Here's the worked-out example. Let's hope this doesn't show a one-day calorie count!

```
      (1)(1)(1)
      4, 2 9 0
      6, 6 1 9
      7, 4 0 4
  +   3, 6 5 2
  ─────────────
    2 1, 9 6 5
```

Try the sample problems.

PROBLEMS

Add the following columns of numbers with pencil and paper.

1.	4 3 0	2.	1, 0 2 1
	7 9		6, 1 4 8
	3 2 9		6, 6 0 7
	+ 9 1 2		+ 5, 2 8 1

3.   6 9 9
     2 5 9
     3 1 1
   + 8 3 4

4.   5, 0 4 9
     1, 0 1 8
       4 3 3
   + 7, 3 4 8

ANSWERS

1. 1,750   **3.** 2,103

2. 19,057   **4.** 13,848

WHAT HAVE YOU LEARNED?

1. Short addition: The numbers from 1 through 9 can be added on your fingers. However, to increase your speed, you should memorize the addition table for numbers below 10.
2. Zero added to any number leaves the number unchanged: 142 + 0 = 142.
3. The order in which you add two numbers does not change the sum: 16 + 8 = 8 + 16.
4. Breaking numbers down: The numbers 0 through 9 are digits. Larger numbers are constructed with a combination of digits. Numbers can be broken down into sums of digits and recombined by using the addition table: 15 + 4 = 10 + (5 + 4) = 10 + 9 = 19.
5. Long addition: To add two numbers using long addition, stack one number on top of the other and draw a line underneath. Beginning with the right-most column of digits, add each column, placing the answer below the line. If a column adds to 10 or more, write the ones digit under the line and add the tens digit to the next column on the left.
6. Addition by columns: When adding columns of numbers, proceed as with long addition, stacking the numbers on top of each other and drawing a line underneath. Then add each column, beginning at the right and carrying when necessary. If you encounter a column

that requires you to add numbers greater than those you memorized from the table, use a scratch pad.

If you've made it through this chapter alive and practiced the basics enough to perform long addition confidently, you have mastered the first major step in mathematics. If you had any difficulty, review the chapter and try the sample problems again.

CHAPTER 5

SUBTRACTION

In this chapter you will augment your mathematical arsenal with subtraction of whole positive numbers (and zero). Later chapters will cover addition and subtraction of other kinds of numbers: negative numbers, fractions, and decimals. If you are confident of your subtraction skills, you might consider skimming this chapter. But before going on to Chapter 6, do the subtraction problems at the end of each section. If you have any trouble, review the appropriate section and try the problems again.

SOME WAYS TO THINK OF SUBTRACTION

If addition is defined as joining or combining items or numbers, then subtraction can be thought of as taking away or deducting items or numbers.

Using groups of things can show subtraction clearly. If you have 5 dollars and subtract 2, you take away 2 dollars from the group of 5 and leave 3. This raises an interesting problem. How do you subtract 5 dollars from 3 dollars? What does negative 2 dollars mean? Basically, it means you're in debt, but you haven't received the collection letter yet.

Think again of the straight number line. When you *add* two positive numbers, you move to the right the number of units indicated by the first number; then you move right again the number of units indicated by the second number.

When you *subtract* a number, you move left on the number line. So, $3 - 5 = (-2)$. How do you get that minus 2? First, move three units *to the right* on the line to end up over the 3. Then, move five units *to the left* to end at the -2. This explains the negative 2 dollars.

SHORT SUBTRACTION

The shortcuts for subtraction are the same as those for addition. Of course, you can subtract by counting on your fingers. Simply reverse the addition process; count down instead of up as you touch your fingers. To subtract 3 from 16, begin with 16, and, as you touch 3 consecutive fingers, count down—15, 14, 13. The answer is 13. As with addition, the drawbacks to this procedure are slowness and the limited number of fingers and toes you have. So it is helpful to memorize the answers, or differences, for the subtraction of numbers less than 18.

You may recall that, in addition, the order of the numbers doesn't matter: $A + B = B + A$, or $7 + 5 = 5 + 7$. With subtraction, however, the order in which you write the numbers is important; 5 minus 1 is 4, but 1 minus 5 is (-4). To use the subtraction table (Table 5.1), take the first number from the left side (5) and the second number from the top (1). To find the answer to 5 minus 1, locate 5 at the left and then read across that row until you are below the 1 at the top. Here you find the number 4. So, $5 - 1 = 4$. Table 5.1 shows answers for the subtraction of the numbers 0 through 9 from the numbers 1

Table 5.1 Subtraction Table

−	1	2	3	4	5	6	7	8	9
1	0								
2	1	0							
3	2	1	0						
4	3	2	1	0					
5	4	3	2	1	0				
6	5	4	3	2	1	0			
7	6	5	4	3	2	1	0		
8	7	6	5	4	3	2	1	0	
9	8	7	6	5	4	3	2	1	0
10	9	8	7	6	5	4	3	2	1
11		9	8	7	6	5	4	3	2
12			9	8	7	6	5	4	3
13				9	8	7	6	5	4
14					9	8	7	6	5
15						9	8	7	6
16							9	8	7
17								9	8
18									9

through 18 (subtractions that yield numbers greater than 10 and less than 0 are not included). These are the subtraction answers you need to master the next section—long subtraction.

As with addition, the key to mastering subtraction is practice, and flash cards are helpful. Knowing as much as you do now, try some problems.

PROBLEMS

Answer each short-subtraction problem. Use Table 5.1 or a calculator to check your answers.

1. 9 − 2 = ?   **6.** 6 − 0 = ?

2. 4 − 4 = ?   **7.** 17 − 9 = ?

3. 15 − 8 = ?   **8.** 13 − 4 = ?

4. 11 − 9 = ?   **9.** 9 − 4 = ?

5. 18 − 9 = ?   **10.** 14 − 7 = ?

ANSWERS

1. 9 − 2 = 7   **6.** 6 − 0 = 6

2. 4 − 4 = 0   **7.** 17 − 9 = 8

3. 15 − 8 = 7   **8.** 13 − 4 = 9

4. 11 − 9 = 2   **9.** 9 − 4 = 5

5. 18 − 9 = 9   **10.** 14 − 7 = 7

LONG SUBTRACTION

Long subtraction is very much like long addition, except that you use a process of *borrowing* instead of *carrying*. Suppose you have 14 eggs and your niece uses 3 for an omelet. Do you have enough eggs left to make Lizette's 12-egg quiche?

First set up the problem as you did for long addition, but make sure that the greater number is on top and the lesser number is on the bottom.

$$\begin{array}{r} 1\ 4 \\ -\quad 3 \\ \hline \end{array}$$

Begin at the right with the digits in the ones place, the 4 and the 3. Subtract 3 from 4, and get 1 as the answer. Write 1 below the line, under the 4 and the 3.

$$\begin{array}{r} 1\ 4 \\ -\quad 3 \\ \hline 1 \end{array}$$

Now move left. Here you have a 1 in the tens place and nothing below it. This is not a problem; you're subtracting nothing. Imagine 0 as the bottom number, and subtract it from 1, which leaves 1. Write the 1 below the line.

$$\begin{array}{r} 1\ 4 \\ -\quad 3 \\ \hline 1\ 1 \end{array}$$

So 14 minus 3 is equal to 11, and you'll need 1 more egg to make the quiche. Now suppose Lizette herself started out with 36 eggs; how many will she have left to try other recipes? Note that this example works the same way.

$$\begin{array}{r} 3\ 6 \\ -\ 1\ 2 \\ \hline 2\ 4 \end{array}$$

So Lizette will have 24 eggs left. So far, you have not needed to borrow anything because, in each column, you've subtracted a lesser digit (or nothing) from a greater digit. But now consider a pet shop that had 42 puppies and sold 29 one weekend. How many puppies were left? Set up the example as before.

$$\begin{array}{r} 4\ 2 \\ -\ 2\ 9 \\ \hline \end{array}$$

You want to subtract 9 from 2, but this would result in a negative number, so you need to *borrow*. You are going to *borrow* 10 from the tens place (at the left) and add it to the 2. This will give you 12, and you can subtract 9 from 12. "Wait!" you say. "What do you mean, borrow 10 from the tens place?" Simple; recall that the 4 in the tens place actually means 40. Changing the 4 to a 3 is really subtracting 10 from 40, giving 30. This allows you to change the 2 to a 12 (add the borrowed 10 to 2).

$$\begin{array}{r} 3(12) \\ \cancel{4}\ \cancel{2} \\ -\ 2\ 9 \\ \hline \end{array}$$

Now you can subtract 9 from 12, resulting in 3.

$$\begin{array}{r} 3(12) \\ \cancel{4}\ \cancel{2} \\ -\ 2\ 9 \\ \hline 3 \end{array}$$

Now move to the tens place and subtract 2 from 3 to get 1. Write the 1 below the line.

$$\begin{array}{r} 3(12) \\ \cancel{4}\ \cancel{2} \\ -\ 2\ 9 \\ \hline 1\ 3 \end{array}$$

The pet shop has 13 puppies left. But how did this problem actually work? To answer this, break the problem apart. Rewrite the original numbers 42 and 29 as (40 + 2) and (20 + 9). Now rewrite the problem.

$$\begin{array}{r}(40+\ 2)\\-(20+\ 9)\\\hline\end{array}$$

To subtract 9 from 2, borrow 10 from 40. This will decrease the 40 to 30 and will increase the 2 to 12.

$$\begin{array}{r}(30+12)\\-(20+\ 9)\\\hline\end{array}$$

Subtract 9 from 12 to get 3 and write the 3 below the line.

$$\begin{array}{r}(30+12)\\-(20+\ 9)\\\hline 3\end{array}$$

Next subtract 20 from 30, and write 10 below the line.

$$\begin{array}{r}(30+12)\\-(20+\ 9)\\\hline 10+\ 3\end{array}$$

Now add 10 and 3 to produce the 13 you want. This expanded illustration shows why the method works, but it's slow and cumbersome. Our *borrowing* shorthand is a handy shortcut.

Rule
(Borrowing) In long subtraction, if you are subtracting a greater digit from a lesser digit, increase the lesser digit by 10, move one place to the left in the same row, and decrease that digit by 1.

Let's say you're driving 391 miles to pick up your neighbors' kid at camp. (We'll say your neighbors broke their legs in a romantic skiing accident, so they can't drive; besides, you owe them a favor— they've had you over for dinner *twice,* and you've never returned the invitation.) You left late and have gone only 42 miles when you stop for lunch. How many more miles do you have to drive? Set up the example.

$$
\begin{array}{r}
3\ 9\ 1 \\
-\ \ 4\ 2 \\
\hline
\end{array}
$$

Begin with the right column. You can't subtract 2 from 1 and get a positive number, so borrow 10 from the tens place. Go to the next column to the left, borrow 1 from the 9 (really taking 10 from 90), and make the 9 an 8 by drawing a line through the 9 and writing an 8 above it. Add 10 to the 1 in the right column to make it 11. You can cross out the 1 and write 11 above it or write a small 1 to the left of the 1 that you're increasing by 10 to show that it's now 11.

$$
\begin{array}{cc}
\begin{array}{r}
8\ 11 \\
3\ \not{9}\ \not{1} \\
-\ \ 4\ 2 \\
\hline
\end{array}
& \text{or}
\begin{array}{r}
8 \\
3\ \not{9}\ 11 \\
-\ \ 4\ 2 \\
\hline
\end{array}
\end{array}
$$

Now you can do the subtraction.

$$
\begin{array}{r}
8 \\
3\ \not{9}\ 11 \\
-\ \ 4\ 2 \\
\hline
3\ 4\ 9 \\
\end{array}
$$

You have 349 miles to drive before you can pick up the neighbors' child. Now you and young Charlie, who turns out not to be such a bad kid after all, are going to drive 604 miles to the ocean. At the end of the first day, you've gone 388 miles. How much farther will you have to drive on the second day?

$$
\begin{array}{r}
6\ 0\ 4 \\
-\ 3\ 8\ 8 \\
\hline
\end{array}
$$

But what do you do when you go to the next digit to the left in order to borrow a 1, and find that that digit is 0. Don't panic. Simply go across the zero one more place or column to the left, and borrow from there.

In our example, you move one place to the left of 4, and find 0. So you go one more place to the left, to 6, and borrow 1 by reducing 6 to 5. Write the borrowed 1 as a 10 above the zero.

Now borrow from the 10, making it a 9, whereas the 4 becomes 14.

$$
\begin{array}{r}
9 \\
5 \,\cancel{1}\cancel{0} \\
\cancel{6}\ \ \cancel{0}14 \\
-\ 3\ \ 8\ \ 8 \\
\hline
\end{array}
$$

You can now complete the subtraction.

$$
\begin{array}{r}
9 \\
5 \,\cancel{1}\cancel{0} \\
\cancel{6}\ \ \cancel{0}14 \\
-\ 3\ \ 8\ \ 8 \\
\hline
2\ \ 1\ \ 6
\end{array}
$$

You have 216 miles to drive on the second day (unless Charlie has a license, in which case you're home free; this is unlikely, however, since he's only eight). Long subtraction enables you to subtract any positive whole number from any other positive whole number. If you are rusty with long subtraction or just learning it, you should practice. Long subtraction is one of the keys to managing your money—balancing your checking account (see Chapter 20), doing your taxes, and so on. Although you will most likely use a calculator to do long subtraction, it's good to know how to do it, in case you ever find yourself without a calculator. If you want to check the accuracy of your subtraction, simply add your answer to the number just above the line to yield the top number. This is an excellent method for reviewing your subtraction since it relies on addition, which is easier for most of us. Try some problems.

PROBLEMS

1.	1 4 − 1 1	**6.**	1 8 − 1 5	**11.**	8 3 4 − 4 4 5

1. 1 4
  − 1 1

2. 2 8
  − 1 8

3. 6 1
  − 4 1

4. 9 9
  − 3 8

5. 4 7
  − 2 3

6. 1 8
  − 1 5

7. 2 9
  − 1 9

8. 4 8
  − 3 9

9. 3 1
  − 2 6

10. 9 0
  − 7 7

11. 8 3 4
  − 4 4 5

12. 5 2 1
  − 3 2 9

13. 4 3 0
  − 1 0 8

14. 9 0 1
  − 8 7 9

15. 6 0 3
  − 4 1 1

ANSWERS

1. 1 4
  − 1 1
  ——
  3

2. 2 8
  − 1 8
  ——
  1 0

3. 6 1
  − 4 1
  ——
  2 0

4. 9 9
  − 3 8
  ——
  6 1

5. 4 7
  − 2 3
  ——
  2 4

6. 1 8
  − 1 5
  ——
  3

7. 2 9
  − 1 9
  ——
  1 0

8.   3
  4̸18
  − 3 9
  ——
  9

9.   2
  3̸11
  − 2 6
  ——
  5

10.   8
  9̸10
  − 7 7
  ——
  1 3

11. 7 12
  8̸ 3̸14
  − 4 4 5
  ——
  3 8 9

12. 4 11
  5̸ 2̸11
  − 3 2 9
  ——
  1 9 2

13.   2
  4 3̸10
  − 1 0 8
  ——
  3 2 2

14. 8 9
  9̸ 10̸11
  − 8 7 9
  ——
  2 2

15. 5
  6̸103
  − 4 1 1
  ——
  1 9 2

WHAT HAVE YOU LEARNED?

Here is a summary of what you should have learned from this chapter.

1. Short subtraction: Memorize Table 5.1, remembering that, in subtraction, the order of the numbers is important: $5 - 4$ is not equal to ($\neq$) $4 - 5$.

2. Long subtraction: To subtract two numbers, write the greater number above the lesser number. Beginning at the right, subtract the bottom digit from the top digit. Continue subtracting digits from right to left.

3. *Borrowing:* In long subtraction, if you are subtracting a greater digit from a lesser digit, increase the lesser digit by 10, move one place to the left in the same row, and decrease that digit by 1:

$$
\begin{array}{r}
3(12) \\
\cancel{4}\ \cancel{2} \\
-\ 2\ 9 \\
\hline
1\ 3
\end{array}
$$

If you've digested the basics of both this chapter and the previous one, and you feel confident about long addition and long subtraction, then you have mastered the second major step in mathematics.

If you had difficulty doing the sample problems at the end of any section, review the appropriate section and try the problems again. A bit of practice, and you'll be on your way.

CHAPTER 6

MULTIPLICATION

Now that you're comfortable with addition and subtraction of positive whole numbers—steel yourself; you're now prepared to tackle the next building block of mathematics—multiplication. You'll master multiplication in slow, easy steps, learning first to multiply only positive whole numbers. Multiplying other numbers (fractions and decimals) will be covered later, after you're a multiplication expert.

MULTIPLICATION IS THE "SUM" OF ITS PARTS

In this chapter you will see the small $\times$ to indicate the operation of multiplication, but multiplication can also be indicated by a dot ($\cdot$) or by two numbers standing side by side.

$$3 \times A = 3 \cdot A = 3A$$

The answer to an addition problem is called a *sum*, whereas the answer to a multiplication problem is called a *product*. Multiplication is really just repeated addition. If you add 3 plus 3, you are doing a kind of multiplication. Adding 3 to 3 gives the same result as multiplying 3 by 2, or 2×3.

$$3 + 3 = 6, \quad \text{and} \quad 2 \times 3 = 6$$

Adding three 3s ($3 + 3 + 3$) is the same as multiplying 3×3. Adding four 3s is the same as multiplying 3 by 4, or 4×3. In a horizontal multiplication problem (such as 3×2), the first number tells you how many equal groups exist. The second number tells you how many items there are in each of the equal groups. So, *3 × 2* means "add 3 groups of 2." Similarly, *22 × 6* means "add 22 groups, each group containing 6 items." You can see that multiplication is merely shorthand for repeated addition of identical numbers. Multiplication is necessary if you think of the difference between writing 22 × 6 and writing an addition problem that consists of 6 written 22 times. (It could take 22 minutes just to write the example!)

One property of multiplication is that the order of the two numbers can be reversed.

$$3 \times 5 = 15 \quad \text{and} \quad 5 \times 3 = 15$$

This means that 5 groups of 3 items have the same total number of items as 3 groups of 5.

Rule
If A and B are numbers, then A × B = B × A.

SHORT MULTIPLICATION

You need a shortcut for multiplying numbers less than 10. Methods exist for multiplying with your fingers, but by the time you master these techniques, you could have memorized the same results from a multiplication table. Table 6.1 is the multiplication table for numbers less than 10.

Table 6.1 Multiplication Table for Numbers Less Than Ten

×	2	3	4	5	6	7	8	9
2	4	6	8	10	12	14	16	18
3		9	12	15	18	21	24	27
4			16	20	24	28	32	36
5				25	30	35	40	45
6					36	42	48	54
7						49	56	63
8							64	72
9								81

The table does not include 0 or 1. Multiplying 0 with any number gives a result of 0, and multiplying any number with 1 gives an answer that is equal to the first number.

Rule
Any number multiplied by 0 results in an answer (product) of 0.

Rule
The result (product) of any number multiplied by 1 is the original number.

Because the order of the numbers doesn't affect the answer in multiplication (as with addition—remember the "cash-register" property?), you need only half the multiplication table. Also, 10 times any number results in that number moved to the left with a 0 written at the right ($10 \times 5 = 50$).

Rule
To multiply a number by 10, shift all the digits one place to the left and write a 0 at the right.

Table 6.1 contains only 36 answers, and they should be easy to memorize. Shortcuts can help: Multiplying any number by 5 results in a number ending with 5 or 0; multiplying an even number by any number results in an even number. Try the following sample problems without looking at the multiplication table. If you have trouble, practice with the table until you can rattle off its answers.

PROBLEMS

1. $6 \times 4 = ?$ **6.** $7 \times 3 = ?$

2. $8 \times 2 = ?$ **7.** $4 \times 0 = ?$

3. $6 \times 7 = ?$ **8.** $9 \times 8 = ?$

4. $7 \times 8 = ?$ **9.** $10 \times 4 = ?$

5. $9 \times 1 = ?$ **10.** $5 \times 2 = ?$

ANSWERS

1. 6 × 4 = 24

2. 8 × 2 = 16

3. 6 × 7 = 42

4. 7 × 8 = 56

5. 9 × 1 = 9

6. 7 × 3 = 21

7. 4 × 0 = 0

8. 9 × 8 = 72

9. 10 × 4 = 40

10. 5 × 2 = 10

LONG MULTIPLICATION

Long multiplication is no more difficult than long addition or long subtraction. Write one number above the other and then apply short multiplication several times. Let's say you are buying 4 tickets to a George Michael concert, and the tickets cost $27 apiece. How much will you have to pay? Set up the problem.

$$
\begin{array}{r}
2\ 7 \\
\times \quad 4 \\
\hline
\end{array}
$$

Begin by multiplying the number at the bottom times the right-most digit in the top row, or 4 × 7. This gives you 28. Break the 28 into 20 + 8. Write the 8 below the line and the 20 above the 27.

$$
\begin{array}{r}
(2\ 0) \\
2\ 7 \\
\times \quad 4 \\
\hline
8 \\
\end{array}
$$

Now multiply 4 times the 2 on top, or 4 × 2. This gives you 8. To this 8, add the tens digit (2) that you carried; 8 + 2 is 10. Write the 10 below the line, to the left of the 8. Note that you don't need to do anything with the 0 above the 7.

$$
\begin{array}{r}
(2\ 0) \\
2\ 7 \\
\times \quad 4 \\
\hline
1\ 0\ 8
\end{array}
$$

You will need $108 to buy your 4 tickets. From now on, you won't need to write the 0 above the 7. How does this method work? We will demonstrate by writing the problem in a kind of longhand. Write 4 × 27 this way.

$$4 \times 27 = 4 \times (20 + 7)$$

Now multiply both the 7 and the 20 separately by 4.

$$(4 \times 20) + (4 \times 7) = 80 + 28$$

Then add 80 and 28 to get 108. You've broken down the problem into a series of steps involving numbers less than 10 and then added the results. The regular method consists of carrying to the left, and is less involved than the above example.

Suppose you are taking part in an intergalactic relay race. Each of the 7 legs of the race is 640,190 miles long. How long is the race? (This is important; you need to know how much Zoz space fuel to buy!) Just set up 7 × 640,190.

$$
\begin{array}{r}
6\ 4\ 0,1\ 9\ 0 \\
\times \qquad\quad 7 \\
\hline
\end{array}
$$

But wait! How do you multiply across zeros? Don't let this throw you. Recall that multiplying any number with 0 equals 0. Simply proceed, adding any carried digits above the 0; then move to the next digit to the left. Take a look at the worked-out example.

$$
\begin{array}{r}
(2)\quad (1)(6) \\
6\ 4\ 0,1\ 9\ 0 \\
\times \qquad\quad 7 \\
\hline
4,4\ 8\ 1,3\ 3\ 0
\end{array}
$$

The relay race is 4,481,330 miles long. (Better buy enough Zoz fuel!) Come back to Earth with some sample problems.

PROBLEMS

1. $77 \times 6 = ?$

2. $183 \times 7 = ?$

3. $2,845 \times 3 = ?$

4. $2,009 \times 4 = ?$

5. $2,654 \times 10 = ?$

6. $102 \times 5 = ?$

7. $991 \times 8 = ?$

8. $7 \times 1,011 = ?$

9. $2 \times 909 = ?$

10. $5 \times 999 = ?$

ANSWERS

1.   (4)
```
    7 7
×     6
  4 6 2
```

2.   (5)(2)
```
    1 8 3
×       7
 1, 2 8 1
```

3. (2)(1)(1)
```
   2, 8 4 5
×        3
   8, 5 3 5
```

4.   (3)
```
   2, 0 0 9
×        4
   8, 0 3 6
```

5. $2,654 \times 10 = 26,540$

6.   (1)
```
    1 0 2
×       5
    5 1 0
```

7.   (7)
```
    9 9 1
×       8
  7, 9 2 8
```

8.   1, 0 1 1
```
×        7
   7, 0 7 7
```

9.   (1)
```
    9 0 9
×       2
  1, 8 1 8
```

10.   (4)(4)
```
    9 9 9
×       5
  4, 9 9 5
```

Now suppose your daughter wants help with her homework one day and asks you how to multiply 24×67. This is actually easy. Just repeat the multiplication process you already know, but watch what happens below the line.

First multiply 4 times 7 to get 28. Write the 8 and carry the 2.

```
        (2)
        6 7
      × 2 4
      ───────
          8
```

Now multiply 4 times 6 to get 24, and add the 2 to get 26. Write the 26 below the line.

```
          (2)
          6 7
      ×   2 4
      ─────────
        2 6 8
```

The next step is to multiply 2 (from 24) times 7 at the top to get 14. Write the 4 below the 268, but shift one digit to the left; write the 4 below the 6. (You're starting the second line of numbers, as you always will in long multiplication, at a position directly below the digit you're multiplying by—in this case, the 2 in 24.) Now carry the 1. Because you've already carried once and placed a 2 above the 6, place the carried 1 above that 2.

```
          (1)
          (2)
          6 7
      ×   2 4
      ─────────
        2 6 8
        4
```

Now multiply 2 times 6 to get 12, and add the carried 1 to get 13. Write the 13 below the 268 and to the left of the 4.

```
          (1)
          (2)
          6 7
      ×   2 4
      ─────────
        2 6 8
      1 3 4
```

Now add the two sets of numbers that are below the line, 268 + 134. Remember to carry, when necessary, for addition.

$$
\begin{array}{r}
(1) \quad \text{(carried from } 2\times7=14) \\
(2) \quad \text{(carried from } 4\times7=28) \\
6\ 7 \\
\times \quad 2\ 4 \\
\hline
\end{array}
$$

(carried from 6+4=10) (1)

$$
\begin{array}{r}
2\ 6\ 8 \\
+\ 1\ 3\ 4 \\
\hline
1,6\ 0\ 8
\end{array}
$$

The answer is 1,608 (and your daughter will be amazed).

If you have any uncertainty about the example, review it slowly, step by step. This example shows how even the most complex multiplication can be carried out.

Now look at a worked-out example for 485 × 692. Notice the alignment of the three partial answers (or partial products). Just remember to move each partial answer one place to the left (beginning below the digit you are multiplying by).

$$
\begin{array}{r}
(3) \\
(7)(1) \\
(4)(1) \\
6\ 9\ 2 \\
\times \quad 4\ 8\ 5 \\
\hline
\end{array}
$$

(1)(1)

(1) 3 4 6 0   ← multiplying by 5 in the ones place

(1) 5 5 3 6   ← multiplying by 8 in the tens place

+ 2 7 6 8   ← multiplying by 4 in the hundreds place

$$
\overline{3\ 3\ 5,6\ 2\ 0}
$$

Although long multiplication is a lengthy procedure, it is based on the repetition of simple ideas. As FDR said to a worried nation facing greater difficulties than the following sample problems, you have nothing to fear but fear itself. Just remember to work carefully through each step.

PROBLEMS

Use long multiplication to solve each problem.

1. 4 3 × 5 4	**6.** 3 5 × 5 5	**11.** 9 1 0 × 2 9 7
2. 7 9 × 4 1	**7.** 6 3 1 × 4 1 2	**12.** 7 3 2 × 4 5 4
3. 8 0 × 9 9	**8.** 4 9 7 × 9 3 6	**13.** 5, 7 7 1 × 1, 0 2 1
4. 6 2 × 1 7	**9.** 8 8 1 × 1 0 1	**14.** 3, 2 0 5 × 8, 9 7 6
5. 9 7 × 9 0	**10.** 1 0 8 × 3 8 5	**15.** 4, 9 7 0 × 4, 9 6 2

ANSWERS

1.
```
      (1)
      (1)
      4 3
×     5 4
     (1)
    1 7 2
  + 2 1 5
  2,3 2 2
```

2.
```
      (3)
      7 9
×     4 1
   (1) 7 9
  + 3 1 6
  3,2 3 9
```

3.
```
      8 0
×     9 9
    7 2 0
  + 7 2 0
  7,9 2 0
```

4.
```
        (1)
        6 2
×       1 7
  (1) 4 3 4
  +     6 2
  1,0 5 4
```

5.
```
      (6)
      9 7
×     9 0
      0 0
  + 8 7 3
  8,7 3 0
```

6.
```
      (2)
      (2)
      3 5
×     5 5
     (1)
    1 7 5
  + 1 7 5
  1,9 2 5
```

7.
```
        (1)
      6 3 1
  ×   4 1 2
  1 2 6 2
    6 3 1
+ 2 5 2 4
2 5 9,9 7 2
```

8.
```
      (8)(6)
      (2)(2)
      (5)(4)
      4 9 7
    × 9 3 6
      (2)
  (1) 2 9 8 2
    1 4 9 1
+ 4 4 7 3
4 6 5,1 9 2
```

9.
```
      8 8 1
  ×   1 0 1
      8 8 1
    0 0 0
+ 8 8 1
8 8,9 8 1
```

10.
```
        (2)
        (6)
        (4)
      1 0 8
    × 3 8 5
  (1) 5 4 0
  (1) 8 6 4
+ 3 2 4
4 1,5 8 0
```

11.
```
      9 1 0
    × 2 9 7
      (1)
  (1) 6 3 7 0
  (1) 8 1 9 0
+ 1 8 2 0
2 7 0,2 7 0
```

12.
```
        (1)
        (1)(1)
        (1)
      7 3 2
    × 4 5 4
      (2)
  (1) 2 9 2 8
  (1) 3 6 6 0
+ 2 9 2 8
3 3 2,3 2 8
```

13.
```
        (1)(1)
      5,7 7 1
    × 1,0 2 1
        (1)
  (1) 5 7 7 1
  1 1 5 4 2
  0 0 0 0
+ 5 7 7 1
5,8 9 2,1 9 1
```

14.
```
        (1)  (4)
        (1)  (4)
        (1)  (3)
        (1)  (3)
      3,2 0 5
    × 8,9 7 6
        (1)(1)
  (1) 1 9 2 3 0
  (1) 2 2 4 3 5
    2 8 8 4 5
+ 2 5 6 4 0
2 8,7 6 8,0 8 0
```

15.
```
        (3)(2)
        (8)(6)
        (5)(4)
        (1)(1)
      4,9 7 0
    × 4,9 6 2
        (1)
  (2)(2) 9 9 4 0
  (1) 2 9 8 2 0
  (1) 4 4 7 3 0
+ 1 9 8 8 0
2 4,6 6 1,1 4 0
```

WHAT HAVE YOU LEARNED?

1. Multiplication can be understood as repeated addition of equal groups: $4 + 4 + 4 = 3 \times 4$, or 12.
2. Rule: If A and B are numbers, then $A \times B = B \times A$: $7 \times 8 = 8 \times 7$ or 56.
3. Rule: Any number multiplied by 0 results in 0 as an answer: $8 \times 0 = 0$.
4. Rule: Multiplying any number by 1 results in a product that is equal to the first number: $23 \times 1 = 23$.
5. Rule: To multiply a number by 10, shift all the digits one place to the left and write a 0 at the right: $6 \times 10 = 60$.
6. Short multiplication: Table 6.1 should be memorized.
7. Long multiplication: Stack one number above the other number and draw a line below them. Begin with the bottom-right digit and multiply this digit across the top number, carrying where necessary. Write the answer below the line. Repeat this process with the next bottom digit to the left, shifting the answer one place to the left. When you have multiplied all the digits of the top number by all the digits of the bottom number, add the partial answers to find the final answer or product:

```
        (2)
        (7)
        7 8 1
      × 3 1 9
      ─────────
      (1)
      7 0 2 9      ←multiplying by 9 in the ones place
  (1) 7 8 1        ←multiplying by 1 in the tens place
+ 2 3 4 3          ←multiplying by 3 in the hundreds place
  ─────────
  2 4 9,1 3 9
```

Ready for division? Of course you are! And all you have to do to begin is turn the page.

CHAPTER 7

DIVISION

It is now time to slay the dragon of arithmetic: division. This will complete your working knowledge of arithmetic's four operations: addition, subtraction, multiplication, and division. As in previous chapters, this chapter will ask you to work only with positive whole numbers.

THE "DIFFERENCE" BETWEEN MULTIPLICATION AND DIVISION

Division is quite different from multiplication. When dividing, the order of the numbers *does* make a difference. The answer to a division problem is called a quotient. Division may seem intimidating at first, but it will become your invaluable tool in solving problems.

You can write a division problem in several ways. Here's how you can write 6 divided by 2.

$$6 \div 2 \quad \text{or} \quad \frac{6}{2} \quad \text{or} \quad {}^{6}\!/_{2} \quad \text{or} \quad 2\overline{)6}$$

Whereas multiplying combines groups of equal numbers, dividing breaks numbers apart—into groups of equal numbers or into a number of equal groups. This is not as complicated as it seems. Suppose you have 6 writing instruments at your desk: 3 pens and 3 pencils. You

59

have 2 groups of 3. You can show that this way: 6 ÷ 2 = 3. Think of the pens and pencils as small rectangles.

6 items
2 groups

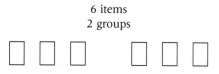

3 items in each group

This example shows you that you have 2 groups that each contain 3 items. In this case, the number you are dividing by (2) is the number of groups, and the quotient (3) is the number of items in each group.

Now suppose that 2 of the 6 writing tools write in blue, 2 in green, and 2 in red. Here's the question: How many equal groups are there? In this case, the number you are dividing by is the number of items in each of the equal groups, or 2. So, 6 ÷ 2 = 3. The quotient tells you that you have 3 equal groups. Look at our rectangular pens and pencils again, labeled for writing color.

6 items
2 items in each group

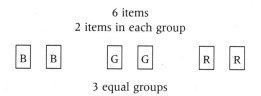

3 equal groups

In division it is important to know what the number you are dividing by means. If the number you are dividing by is the number of equal groups, the quotient will tell you how many items are in each of the groups. If the number you are dividing by is the number of items in each of the equal groups, the quotient will tell you how many of those groups you have.

SHORT DIVISION

Division relates to multiplication in much the same way that subtraction relates to addition. Because division is the inverse operation of multiplication, the multiplication table is the court of first resort to solve division problems. Suppose you have a $42 check at a restaurant, and 6 people are at the table. When you're looking to divide 42 by 6, what you are *really* looking for is a number that will give 42 as an answer when multiplied by 6. Think of it this way: 6 × ? = 42. Find these numbers in Table 7.1, the multiplication table.

Look down the table until you come to the 6 at the far left. Track across that row until you reach 42. Track up that column to find 7 at the top of the table. So 7 is the number you are after—6 × 7 = 42, or 42 ÷ 6 = 7. If you split the check evenly, each person at the table owes exactly $7. However, not all division is so neat.

Suppose you and five cousins have breakfast to decide on a Valentine's gift for your grandparents. The breakfast comes to $45. How do you divide the bill equally?

Remember that each step in the math stairway is built on the preceding step. Use multiplication to help with division. If you can't find the number you are dividing in the multiplication table, find the two products between which the number you are dividing lies. Think again of the question—a bill at a restaurant for $45, and 6 people are at the table. Set up the problem.

$$6\overline{)45}$$

The first question to ask is how many times 6 will go into 4 (since 4 is the first digit under the division housing). Well, 6 doesn't go into 4 even once. Move right to include the 5, and ask how many times 6 will go into 45. You want to find the greatest number you can multiply

Table 7.1 Multiplication Table for Numbers Less Than Ten

×	2	3	4	5	6	7	8	9
2	4	6	8	10	12	14	16	18
3		9	12	15	18	21	24	27
4			16	20	24	28	32	36
5				25	30	35	40	45
6					36	42	48	54
7						49	56	63
8							64	72
9								81

by 6 without exceeding 45. The multiplication table tells you: 7. Write the 7 above the line over the 5, because 6 goes into 45, not into 4.

$$
\begin{array}{r}
7 \\
6\overline{\smash{\big)}\,4\ 5}
\end{array}
$$

Multiply 6 × 7 and write the result, 42. Subtract.

$$
\begin{array}{r}
7 \\
6\overline{\smash{\big)}\,4\ 5} \\
4\ 2 \\
\hline
3
\end{array}
$$

The 3 is less than 6, so you're finished. You can divide no further. To say that 45 divided by 6 gives a result of 7 with a remainder of 3 is the same as saying that 45 can be separated into 7 groups of 6 plus one more group with only 3. We do not have enough numbers to form an eighth group containing 6. For this we would have to divide 6 into 48. Hence, the remainder to a division problem tells us how many we have left over after separating the number being divided into groups. How do we write the remainder of 3 in our answer? We form a fraction with the top number of 3 and a bottom number of 6. This fraction tells us that we have 3 out of 6 numbers necessary to make an eighth group. Therefore, our final answer becomes 7 + $3/_6$ or 7$3/_6$, which equals 7½. So each cousin owes $7.50.

Now your Aunt Elsa asks you to find out how much the 7 cousins will each owe for a 75th birthday celebration for Uncle Ozzie if the total cost is $5,613. Set up the problem.

$$
7\overline{\smash{\big)}\,5{,}6\ 1\ 3}
$$

Does 7 go into 5? No. Move one place to the right. Does 7 go into 56? Yes, 7 goes into 56 exactly 8 times: 7 × 8 = 56. Write 56 below the original 56. Subtract 56 from 56.

$$
\begin{array}{r}
8 \\
7\overline{\smash{\big)}\,5{,}6\ 1\ 3} \\
5\ 6 \\
\hline
0
\end{array}
$$

Write an x under the 1. Then copy the 1 under the line and to the right of the zero. We call this procedure "bringing down" the one. An x is placed under the original 1 to remind us later on that we have carried it down. Now we repeat the division process using the number under the line.

```
          8
    ┌──────────
7   │  5,6 1 3
    │  5 6 x
    ├──────────
       0 1
```

How many times will 7 go into 1? None. Write a zero in the answer to the right of 8 and bring down the last digit, the 3.

```
          8 0
    ┌──────────
7   │  5,6 1 3
    │  5 6 x x
    ├──────────
       0 1 3
```

How many times will 7 go into 13? Once. Write 1 to the right of 0 in the answer. Multiply 1 times 7, resulting in 7, and write 7 below the bottom 3. Subtract.

```
          8 0 1
    ┌──────────
7   │  5,6 1 3
    │  5 6 x x
    ├──────────
       0 1 3
           7
    ├──────────
          6   (remainder)
```

The remainder is 6. You have no more digits to bring down. So, each of the 7 cousins owes $801, plus change. (If you're going to feud about the $6/7$ of a dollar, you might as well not bother coming to Uncle Ozzie's party anyway.)

This problem illustrates an important point. If after bringing a digit down, the resulting number is too small for the number you are dividing by to go into it at least once, write a 0 in the answer above the digit you brought down; then bring down the next digit. Look again at steps 2 and 3 of the problem you've just done.

```
            8
      ┌─────────
    7 │ 5,6 1 3
        5 6 x
      ─────────
          0 1
```

Bring down the 1. 7 does not go into 1.

```
           8 0
      ┌─────────
    7 │ 5,6 1 3
        5 6 x x
      ─────────
          0 1 3
```

Write 0 in the quotient above the 1. Bring down the next digit, 3.

Try some short-division problems.

PROBLEMS

1. $201/5 = ?$

2. $1,892/4 = ?$

3. $6,321/7 = ?$

4. $7,047/2 = ?$

5. $99/9 = ?$

6. $339/3 = ?$

7. $10,111/9 = ?$

8. $661/8 = ?$

9. $7,852/6 = ?$

10. $1,000/9 = ?$

ANSWERS

```
1.        4 0
      ┌─────────
    5 │ 2 0 1
        2 0 x
      ─────────
          0 1  (remainder)
```

```
2.        4 7 3
      ┌─────────
    4 │ 1,8 9 2
        1 6 x x
      ─────────
          2 9
          2 8
        ─────────
            1 2
            1 2
          ─────────
              0  (remainder)
```

```
3.        9 0 3
      ┌─────────
    7 │ 6,3 2 1
        6 3 x x
      ─────────
          0 2 1
            2 1
          ─────────
              0  (remainder)
```

```
4.        3,5 2 3
      ┌─────────
    2 │ 7,0 4 7
        6 x x x
      ─────────
          1 0
          1 0
        ─────────
            0 4
            4
          ─────────
            0 7
            6
          ─────────
              1  (remainder)
```

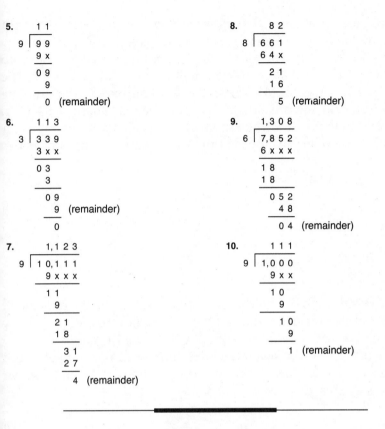

```
5.      1 1                    8.       8 2
    9 | 9 9                        8 | 6 6 1
        9 x                            6 4 x
       ───                            ─────
        0 9                            2 1
        9                              1 6
       ───                            ─────
        0  (remainder)                 5  (remainder)


6.      1 1 3                  9.      1,3 0 8
    3 | 3 3 9                        6 | 7,8 5 2
        3 x x                            6 x x x
       ─────                            ───────
        0 3                              1 8
        3                                1 8
       ───                              ─────
        0 9                              0 5 2
        9  (remainder)                   4 8
       ───                              ─────
        0                                0 4  (remainder)


7.      1,1 2 3                10.      1 1 1
    9 | 1 0,1 1 1                   9 | 1,0 0 0
        9 x x x                          9 x x
       ───────                          ─────
        1 1                              1 0
        9                                9
       ───                              ───
        2 1                              1 0
        1 8                              9
       ───                              ───
        3 1                              1  (remainder)
        2 7
       ───
        4  (remainder)
```

LONG DIVISION

The next step is to divide by a number that contains more than one digit. Your Uncle Ozzie has to drive 92 miles to pick up a gift for Aunt Elsa. His big old station wagon, still mobile after twenty-some years, gets 17 miles per gallon of gas. How many gallons of gas does Ozzie need to buy? Set up the problem.

$$17 \overline{)\,92}$$

Look at the first digit in the dividend, 9. Will 17 go into 9? No. Move right. Will 17 go into 92? You can't use the multiplication table to decide. Or can you? In a way, using the table can help.

You can use estimation: 17 is close to 20, and 92 is close to 100. Ask how many times 20 will go into 100 (which is like asking how many times 2 goes into 10). The answer is 5. So estimate that 17 goes into 92 close to 5 times. Write 5 above the line and multiply 5 × 17. The answer is 85. Write the 85 below the 92. Fortunately, the estimate was correct, and the answer, 85, is less than 92. Subtract 85 from 92; 7 is the remainder. If the estimate had yielded a number greater than 92, you would have needed to begin again with a number less than 5.

$$
\begin{array}{r}
5 \\
17\,\overline{)\,9\,2} \\
8\,5 \\
\hline
7 \quad \text{(remainder)}
\end{array}
$$

Uncle Ozzie has to buy at least 5 gallons of gas. Given the remainder, he should probably buy 6 gallons. (And given Uncle Ozzie's propensity for getting lost, he should probably buy an extra 10 gallons.)

Let's do the problem again, but this time we'll pretend we didn't think to try the 5 right away. Let's pretend we thought the best number to try was 4. This gives us:

$$
\begin{array}{r}
4 \\
17\,\overline{)\,9\,2}
\end{array}
$$

We multiply 4 times 17 and get 68. We write the 68 below the 92, draw a line, and subtract.

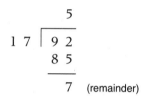

The number we get after subtracting is 24, which is larger than 17. When this occurs, we have guessed a number which is too small. Therefore, the 4 must be increased and the problem restarted. Remember, if the result of our subtraction is larger than the number we are dividing by, our guess was too small.

Now let's progress to a more difficult long division problem. We'll divide 791 by 62.

$$62\,|\,\overline{791}$$

How many times will 62 go into 79? Once.

$$62\,|\,\overline{\begin{matrix}1\\791\\\underline{6\,2}\\1\,7\end{matrix}}$$

We went ahead and multiplied 1 by 62, put the results below the 79, and subtracted. The result of subtraction is 17. Now we carry down the 1 which is next to the 9 in the original problem.

$$62\,|\,\overline{\begin{matrix}1\\791\\\underline{6\,2\,x}\\1\,7\,1\end{matrix}}$$

How many times will 62 go into 171? Frankly, I don't know. We're going to have to guess. We know that 62 is almost 50 and 171 is almost 150. Fifty will go into 150 three times, so 3 is a good number to try. Put 3 down next to the 1 above the line.

$$62\,|\,\overline{\begin{matrix}1\,3\\791\\\underline{6\,2\,x}\\1\,7\,1\end{matrix}}$$

Now multiply 3 times 62 to get 186. Because 186 is greater than 171, we must have chosen too great a number. Three is not the number we want. Cross out the 3 and write 2 above it.

```
                2
            1 3̸
    6 2 | 7 9 1
          6 2 x
        ─────────
          1 7 1
```

Multiply 2 times 62 to get 124 and write the 124 below the 171.

```
                2
            1 3̸
    6 2 | 7 9 1
          6 2 x
        ─────────
          1 7 1
          1 2 4
        ─────────
            4 7
```

Subtracting 124 from 171 we get 47. Therefore, the answer is 12 and 47/62. Remember, we crossed out the 3 in the answer and replaced it with the 2. It's important to cross out our wrong guess so they don't accidentally end up in our final answer.

Remember: If you multiply the dividing number by your estimate and find a number greater than the number you're dividing into, you must reduce your estimate by 1 and try again. Fortunately, you can usually use a calculator to perform division. Try the sample problems and check your estimates with a calculator.

PROBLEMS

1. 1 6 | 4 1 1   6. 3 7 | 6 9 2

2. 9 2 | 1,0 1 1   7. 6 7 | 7 9 3

3. 3 6 1 | 2,9 3 7   8. 8 3 3 | 1 0,8 8 3

4. 6 0 2 | 3,4 8 8   9. 3 9 | 7 9,7 3 5

5. 1 0 0 2 | 9 0,4 1 1   10. 5 7 | 1 1 0,1 0 1

ANSWERS

1.
```
        2 5
  1 6 | 4 1 1
        3 2 x
        ─────
          9 1
          8 0
        ─────
          1 1  (remainder)
```

2.
```
            1 0
  9 2 | 1,0 1 1
        9 2 x
      ───────
          9 1  (remainder)
```

3.
```
                8
  3 6 1 | 2,9 3 7
          2 8 8 8
        ─────────
              4 9  (remainder)
```

4.
```
                5
  6 0 2 | 3,4 8 8
          3 0 1 0
        ─────────
            4 7 8  (remainder)
```

5.
```
                  9 0
  1,0 0 2 | 9 0,4 1 1
            9 0 1 8 x
          ───────────
                2 3 1  (remainder)
```

6.
```
          1 8
  3 7 | 6 9 2
        3 7 x
      ───────
        3 2 2
        2 9 6
      ───────
          2 6  (remainder)
```

7.
```
          1 1
  6 7 | 7 9 3
        6 7 x
      ───────
        1 2 3
          6 7
        ─────
          5 6  (remainder)
```

8.
```
              1 3
  8 3 3 | 1 0,8 8 3
            8 3 3 x
          ─────────
            2 5 5 3
            2 4 9 9
          ─────────
                5 4  (remainder)
```

9.
```
            2 0 4 4
  3 9 | 7 9,7 3 5
        7 8 x x x
      ───────────
          1 7 3
          1 5 6
        ───────
            1 7 5
            1 5 6
          ───────
              1 9  (remainder)
```

10.
```
                1 9 3 1
  5 7 | 1 1 0,1 0 1
        5 7 x x x
      ─────────────
        5 3 1
        5 1 3
      ───────
          1 8 0
          1 7 1
        ───────
              9 1
              5 7
            ─────
              3 4  (remainder)
```

WHAT HAVE YOU LEARNED?

1. Write the number being divided inside the division housing and the number you are dividing by to the left of the housing: $7 \div 2 =$

$$2\ \overline{)\ 7.}$$

2. Estimate to find the maximum number of times the dividing number will go into the appropriate digit(s) of the number being divided. Use your knowledge of the multiplication table, Table 6.1: Because $6 \times 7 = 42$, you know that 6 goes into 43 at least 7 times. Write this estimate above the line as part of your answer above the appropriate digit under the housing.

3. Multiply your estimate by the dividing number and write the answer beneath the appropriate digits.

4. Subtract the multiplied number from the number above it and write the answer below. If no other digits remain under the division housing to be brought down, you are finished.

5. If another digit remains in the number within the housing, bring the digit down and write it to the right of the result of your subtraction.

6. Repeat the division process until all the digits within the housing are used.

You have now covered the four basic operations of arithmetic: addition, subtraction, multiplication, and division. These operations enable you to do all of the operations in the following chapters. If you are uncomfortable with any operation, review Chapters 4 through 7. Practice always helps.

If you are comfortable so far, give yourself a pat on the back. You have learned to conquer the most fundamental steps in mathematics. You are ready to move on.

CHAPTER 8

NEGATIVE NUMBERS

Although you may seldom need to work with negative numbers, you might find some pleasure in seeing that you now know enough math to manipulate them.

The rules for handling negative numbers are very simple. However, you must be careful not to mistake the negative sign for a minus sign. As long as you distinguish between a negative number and the operation of subtraction, you should have no trouble.

ADDING NEGATIVE NUMBERS

Rule
To add two numbers that have the same sign (positive or negative), simply add the numbers and write the common sign.

Suppose you borrowed $493 from your friend Jim on Monday and then $63 more on Thursday. How much do you owe Jim now? Treat the numbers as negative: $(-493) + (-63)$.

$$(-493) + (-63) = \begin{array}{r} (1) \\ 4\ 9\ 3 \\ +\quad 6\ 3 \\ \hline 5\ 5\ 6 \end{array} = (-556)$$

Despite the fat wad of bills in your hand, you actually have -556 dollars, because you owe Jim $556.

To add numbers with opposite signs, use the following rule.

71

Rule

To add two numbers that have opposite signs (one negative and one positive), subtract the lesser number from the greater number and write the sign of the greater number.

Several weeks later, you've paid back all but $3 to Jim. You have $9 in your wallet, but how much money do you actually have? Remember that the $3 you still owe is a negative number: $(+9) + (-3)$. Simply subtract 3 from 9 and write a positive sign $(+)$ in the answer, because the greater number, 9, has a positive sign.

$$(+9) + (-3) = \frac{\begin{array}{r} 9 \\ -\,3 \end{array}}{6} = (+6)$$

You actually have $6. (The other 3 dollars belong to Jim!)

However, if you had $4 and still owed Jim $7, you'd be in the hole. Try it: $(+4) + (-7)$. The digit 7 is greater than the digit 4. So, subtract 4 from 7. Now write a negative sign, because the larger number 7, has a negative sign.

$$(+4) + (-7) = \frac{\begin{array}{r} 7 \\ -\,4 \end{array}}{3} = (-3)$$

You owe $3 that you don't even have … maybe borrowing more isn't the answer.

Try some problems, remembering the difference between plus and minus signs and positive and negative signs.

PROBLEMS

1. $(+4) + (-1) = ?$

2. $(+51) + (-4) = ?$

3. $(-15) + (+2) = ?$

4. $(+142) + (-23) = ?$

5. $(-17) + (-3) = ?$

6. $(-1,330) + (+192) = ?$

7. $(+14) + (-667) = ?$

8. $(+11) + (+21) = ?$

9. $(-197) + (-65) = ?$

10. $0 + (-44) = ?$

ANSWERS

1. (+4) + (−1) = (+3)

2. (+51) + (−4) = (+47)

3. (−15) + (+2) = (−13)

4. (+142) + (−23) = (+119)

5. (−17) + (−3) = (−20)

6. (−1,330) + (+192) = (−1,138)

7. (+14) + (−667) = (−653)

8. (+11) + (+21) = (+32)

9. (−197) + (−65) = (−262)

10. 0 + (−44) = (−44)

SUBTRACTING NEGATIVE NUMBERS

Rule

To subtract when one or both numbers are negative numbers, make the subtraction problem an addition problem by changing the sign of the second number.

You have $4 of your own and $3 that you borrowed from Jim, but Jim tells you to forget about paying back the $3. How much money do you actually have? Here, subtracting a negative is the same as adding a positive: (+4) − (−3). Change the problem to an addition problem: Make the minus sign a plus sign and then change the negative sign in front of the 3 to a positive sign.

$$(+4) - (-3) = (+4) + (+3)$$

By completing a simple addition problem, (+4) + (+3) = (+7), you can see that you actually have $7.

If you borrowed $4 from Jim and he says you don't have to pay back $3 of it, how much do you owe? You can follow the same steps to subtract a negative number from a negative number: (−4) − (−3).

$$(-4) - (-3) = (-4) + (+3) = (-1)$$

You still owe $1.

Try the sample problems.

PROBLEMS

1. $(-21) - (+12) = ?$

2. $(+1,321) - (-299) = ?$

3. $(-11) - (-49) = ?$

4. $(-6,714) - (+7,339) = ?$

5. $(+101) - (-34) = ?$

6. $(+2) - (+166) = ?$

7. $(-43) - (-54) = ?$

8. $(-17) - (+99) = ?$

9. $(-99) - 0 = ?$

10. $0 - (-12) = ?$

ANSWERS

1. $(-21) - (+12) = (-21) + (-12) = (-33)$

2. $(+1,321) - (-299) = (+1,321) + (+299) = (+1,620)$

3. $(-11) - (-49) = (-11) + (+49) = (+38)$

4. $(-6,714) - (+7,339) = (-6,714) + (-7,339) = (-14,053)$

5. $(+101) - (-34) = (+101) + (+34) = (+135)$

6. $(+2) - (+166) = (+2) + (-166) = (-164)$

7. $(-43) - (-54) = (-43) + (+54) = (+11)$

8. $(-17) - (+99) = (-17) + (-99) = (-116)$

9. $(-99) - 0 = (-99) + 0 = (-99)$

10. $0 - (-12) = 0 + (+12) = (+12)$

MULTIPLYING AND DIVIDING NEGATIVE NUMBERS

Multiplying and dividing negative numbers is very easy. Follow two simple rules.

Rule

If the two numbers in a multiplication or division problem have opposite signs, the answer is always negative.

Rule

If the two numbers in a multiplication or division problem have the same sign (either positive or negative), the answer is always positive.

You placed \$2 bets on 6 different horses at the track and lost each race. How much money have you lost? Multiply $(-2) \times (+6)$ to get a negative answer, or $(-2) \times (+6) = (-12)$. You are down \$12. If the signs are the same, the answer is positive: $(-2) \times (-6) = (+12)$.

Division works the same way. Divide: $(+6) \div (-2)$ Carry out the division and write a negative sign: $(+6) \div (-2) = (-3)$. If both numbers were negative, the answer would be positive: $(-6) \div (-2) = (+3)$. Try the sample problems.

PROBLEMS

1. $(+3) \times (-4) = ?$

2. $(+33) \times (+2) = ?$

3. $(-28) \div (-4) = ?$

4. $(-9) \times (-4) = ?$

5. $(+15) \div (+3) = ?$

6. $(-32) \div (+8) = ?$

7. $(-49) \div (-7) = ?$

8. $(-6) \times (+4) = ?$

9. $(-11) \times 0 = ?$

10. $0 \div (-2) = ?$

ANSWERS

1. $(+3) \times (-4) = (-12)$

2. $(+33) \times (+2) = (+66)$

3. $(-28) \div (-4) = (+7)$

4. $(-9) \times (-4) = (+36)$

5. $(+15) \div (+3) = (+5)$

6. $(-32) \div (+8) = (-4)$

7. $(-49) \div (-7) = (+7)$

8. $(-6) \times (+4) = (-24)$

9. $(-11) \times 0 = 0$

10. $0 \div (-2) = 0$

Remember that 0 multiplied by any number results in 0, and 0 divided by any number results in 0.

WHAT HAVE YOU LEARNED?

You have worked with five useful rules for handling problems that have negative numbers. If you have mastered the rules, you have learned to perform computations that involve negative numbers.

1. Rule: To add two numbers that have the same sign (positive o: negative), add the numbers and write the common sign: $(-3) +$ $(-5) = (-8)$.

2. Rule: To add two numbers that have opposite signs (one negative and one positive), subtract the lesser number from the greater num ber and write the sign of the greater number: $(+4) + (-7) = (-3)$

3. Rule: To subtract when one or both numbers are negative numbers make the subtraction problem an addition problem by changing the sign of the second number: $(-4) - (-3) = (-4) + (+3) = (-1)$.

4. Rule: If the two numbers in a multiplication or division problem are of opposite signs, the answer is negative: $(-2) \times (+6) = (-12)$ $(+6) \div (-2) = (-3)$.

5. Rule: If the two numbers in a multiplication or division problem have the same sign (either positive or negative), the answer is al ways positive: $(-2) \times (-6) = (+12)$; $(-6) \div (-2) = (+3)$.

CHAPTER 9

CHANGING ONE

FRACTION

INTO ANOTHER

In this chapter you'll learn how to change one fraction into another fraction of equal value. You might ask: why would we even want to do such a thing? Several good reasons exist. Sometimes it is desirable to change a fraction with large numbers above and below into another fraction with small numbers to make computations easier. This is called "reducing" or "simplifying" fractions. In order to add and subtract fractions we sometimes need to change them. Therefore, changing one fraction into another of equal value is useful.

CHANGING A FRACTION WITH MULTIPLICATION

Each point on the number line can be represented by an infinite number of fractions. Take the point represented by the fraction $\frac{1}{2}$, which is halfway between the whole numbers 0 and 1. This point can also be represented by the following fractions.

$$2/4 \quad 5/10 \quad 55/110 \quad 91/182 \quad 1,011/2,022$$

All the above fractions are equal in value to one another because they all represent the same point on the number line. But you'd certainly rather add $\frac{1}{2}$ and $\frac{1}{2}$ than $1,011/2,022$ and $1,011/2,022$.

By multiplying a fraction's top number and its bottom number by the same number (excluding 0 and 1), you produce a different fraction that has the same value as the original fraction.

Rule

Multiplying the top and bottom numbers of a fraction by the same number (excluding 0 and 1) yields another fraction that has the same value.

You and your friend are given a loaf of bread. You split it down the middle, each taking half. Then you each cut your half into 5 slices to make toast. How much of the loaf is yours to toast? Take the fraction $\frac{1}{2}$ and multiply the top and the bottom by 5.

$$\frac{1}{2} = \frac{1 \times 5}{2 \times 5} = \frac{5}{10}$$

So the fraction $\frac{5}{10}$ is equal to the fraction $\frac{1}{2}$ and you get to toast 5 slices. For one more example, multiply the top and the bottom of $\frac{7}{11}$ by 3.

$$\frac{7}{11} = \frac{7 \times 3}{11 \times 3} = \frac{21}{33}$$

You see that $\frac{7}{11}$ is equal to $\frac{21}{33}$. Try a few sample problems.

PROBLEMS

Multiply the top and bottom numbers of each fraction by the number indicated.

1. $\frac{1}{3}$ by 4

2. $\frac{1}{9}$ by 6

3. $\frac{3}{4}$ by 3

4. $\frac{2}{13}$ by 5

5. $\frac{4}{7}$ by 16

6. $\frac{109}{51}$ by 11

ANSWERS

1. $\frac{1}{3} = \frac{1 \times 4}{3 \times 4} = \frac{4}{12}$

2. $\frac{1}{9} = \frac{1 \times 6}{9 \times 6} = \frac{6}{54}$

3. $\frac{3}{4} = \frac{3 \times 3}{4 \times 3} = \frac{9}{12}$

4. $\frac{2}{13} = \frac{2 \times 5}{13 \times 5} = \frac{10}{65}$

5. $\frac{4}{7} = \frac{4 \times 16}{7 \times 16} = \frac{64}{112}$

6. $\frac{109}{51} = \frac{109 \times 11}{51 \times 11} = \frac{1,199}{561}$

CHANGING A FRACTION WITH DIVISION

You can also *divide* the top and bottom numbers of a fraction by the same number (excluding 0 and 1) to get a fraction of equal value.

Rule
Dividing the top and bottom numbers of a fraction by the same number (excluding 0 and 1) yields another fraction that has the same value.

The procedure for changing fractions by dividing requires close attention. Multiplying two whole numbers always yields a whole number as an answer. However, dividing a whole number by a whole number does not always yield another whole number. You must be careful when choosing the number by which to divide.

Let's begin with $2/4$. Sally has made a batch of pudding and divided it into 4 cups. You eat one, and it is so good that you eat another. How much of the original batch of pudding did you eat? Both the top and bottom number can be divided exactly by the number 2.

$$\frac{2}{4} = \frac{2 \div 2}{4 \div 2} = \frac{1}{2}$$

By eating 2 of the 4 cups, you ate $1/2$ of all the pudding.

You have divided the top number, 2, and the bottom number, 4, by 2 to get the new fraction, $1/2$. Divide the top and bottom of $15/12$ by 3, and see what happens.

$$\frac{15}{12} = \frac{15 \div 3}{12 \div 3} = \frac{5}{4}$$

So $5/4$ is equal to $15/12$. Try the sample problems.

PROBLEMS

Divide the top and bottom numbers of the following fractions by the number indicated.

1. $3/6$ by 3

2. $7/14$ by 7

3. $24/30$ by 6

4. $15/5$ by 5

5. $110/55$ by 11

6. $24/144$ by 12

ANSWERS

1. $\frac{3}{6} = \frac{3 \div 3}{6 \div 3} = \frac{1}{2}$

4. $\frac{15}{5} = \frac{15 \div 5}{5 \div 5} = \frac{3}{1}$

2. $\frac{7}{14} = \frac{7 \div 7}{14 \div 7} = \frac{1}{2}$

5. $\frac{110}{55} = \frac{110 \div 11}{55 \div 11} = \frac{10}{5}$

3. $\frac{24}{30} = \frac{24 \div 6}{30 \div 6} = \frac{4}{5}$

6. $\frac{24}{144} = \frac{24 \div 12}{144 \div 12} = \frac{2}{12}$

BREAKING DOWN NUMBERS INTO PRIMES

Consider the fraction $\frac{1}{2}$. Can you think of a whole number (other than the number 1) by which to divide both the top and the bottom that will produce whole numbers for the top and bottom? No. Any whole number you try to divide by (with the exception of 1) will not divide $\frac{1}{2}$ exactly. The fraction $\frac{1}{2}$ is a special fraction because it is in its lowest terms, or simplest form. Dividing a fraction's top and bottom numbers to get another fraction is called simplifying the fraction.

Definition
A fraction whose top and bottom numbers cannot be divided exactly by the same number to yield a new, simpler fraction is in its lowest terms.

When you simplify fractions to find their lowest terms, there is an easy way to proceed. You can break the fraction's whole numbers apart by a method known as factoring.

Consider the number 4. You know that $2 \times 2 = 4$, and 4 can be broken (or factored) into 2×2. You can't do this with all numbers. The number 7, for example, can't be broken down into two smaller whole numbers multiplied together. The only factors of 7 are 7 and 1, but that does not help you to break down the 7. The number 1 multiplied with any other number does not change the value of that number. Numbers that cannot be broken down into factors are called prime numbers.

Definition
A prime number is a whole number that cannot be broken down into smaller whole numbers multiplied together.

Consider the number 14. Can you break it down further, or is it prime? Any even number can be divided exactly by 2, and 14 is even. So 14 ÷ 2 = 7; 2 × 7 = 14. You've discovered how to break 14 down at least once: 14 = 2 × 7. Try some problems.

PROBLEMS

Break down each number into as many factors as you can.

1. 6   **5.** 49
2. 12   **6.** 54
3. 18   **7.** 81
4. 30   **8.** 121

ANSWERS

1. $6 = 2 \times 3$

2. $12 = 3 \times 4 = 3 \times 2 \times 2$

3. $18 = 2 \times 9 = 2 \times 3 \times 3$

4. $30 = 5 \times 6 = 5 \times 2 \times 3$

5. $49 = 7 \times 7$

6. $54 = 9 \times 6 = 3 \times 3 \times 2 \times 3$

7. $81 = 9 \times 9 = 3 \times 3 \times 3 \times 3$

8. $121 = 11 \times 11$

CANCELLATION

The technique of breaking down numbers, or factoring, is useful in changing fractions. If you break down the fraction's top and bottom numbers, you can see quickly which numbers to divide by in order to simplify the fraction. Instead of laboriously carrying out the division, you can just cancel out the appropriate numbers from the fraction's top and bottom.

Consider the fraction $\frac{7}{14}$. You can't break the top number down any further, but the bottom number breaks into 2 × 7. Rewrite the fraction in factored form.

$$\frac{7}{14} = \frac{7}{2 \times 7}$$

On the right side of the above equation, there is a 7 in both the top number and the bottom number of the fraction. If you divided by 7 you would get a 1 in the top number and a 2 in the bottom number You can just cancel the 7s by drawing lines through them.

$$\frac{7}{14} = \frac{7}{2 \times 7} = \frac{\cancel{7}}{2 \times \cancel{7}} = \frac{1}{2}$$

By factoring the numbers in fractions and then canceling out the common prime numbers, you reduce the fraction to its lowest terms

Rule
Canceling a number from a fraction eliminates common prime factor from both the top and the bottom of the fraction.

Try a few problems.

PROBLEMS

By factoring and canceling, change the following fractions into fractions in their lowes terms.

1. $^{21}/_{7}$

2. $^{44}/_{11}$

3. $^{30}/_{21}$

4. $^{9}/_{99}$

5. $^{55}/_{105}$

6. $^{56}/_{32}$

ANSWERS

1. $\frac{21}{7} = \frac{3 \times 7}{7} = \frac{3 \times \cancel{7}}{\cancel{7}} = \frac{3}{1} = 3$

2. $\frac{44}{11} = \frac{2 \times 2 \times 11}{11} = \frac{2 \times 2 \times \cancel{11}}{\cancel{11}} = \frac{2 \times 2}{1} = 4$

3. $\frac{30}{21} = \frac{5 \times 2 \times 3}{3 \times 7} = \frac{5 \times 2 \times \cancel{3}}{\cancel{3} \times 7} = \frac{5 \times 2}{7} = \frac{10}{7}$

4. $\frac{9}{99} = \frac{3 \times 3}{3 \times 3 \times 11} = \frac{\cancel{3} \times \cancel{3}}{\cancel{3} \times \cancel{3} \times 11} = \frac{1}{11}$

5. $\frac{55}{105} = \frac{5 \times 11}{5 \times 3 \times 7} = \frac{\cancel{5} \times 11}{\cancel{5} \times 3 \times 7} = \frac{11}{3 \times 7} = \frac{11}{21}$

6. $\frac{56}{32} = \frac{7 \times 2 \times 2 \times 2}{2 \times 2 \times 2 \times 2 \times 2} = \frac{7 \times \cancel{2} \times \cancel{2} \times \cancel{2}}{2 \times 2 \times \cancel{2} \times \cancel{2} \times \cancel{2}} = \frac{7}{2 \times 2} = \frac{7}{4}$

WHAT HAVE YOU LEARNED?

You now know a number of useful rules for changing one fraction into another fraction with the same value.

1. Rule: Multiplying or dividing the top and bottom numbers of a fraction by the same number (excluding 0 and 1) yields another fraction with the same value.
2. Definition: A fraction whose top and bottom numbers cannot be divided exactly by the same number to yield a new, simpler fraction is in its lowest terms.
3. Definition: A prime number is a whole number that cannot be broken down into smaller whole numbers multiplied together.
4. Rule: Canceling a number from a fraction eliminates the same number from the top and bottom numbers of the fraction.

Changing fractions is necessary in order to add and subtract fractions. A fraction in its lowest terms is easier to manipulate in addition, subtraction, multiplication, and division.

You now know that you can multiply and divide both the top and the bottom of a fraction by the same number without changing the fraction's value. Does this mean that you can also add or subtract the same number to or from the top and bottom and leave the value unchanged? No. But you'll find out more about this in Chapter 10.

FRACTIONS
MADE EASY

Because fractions are numbers, the four basic operations of arithmetic apply to them. This chapter shows you how to operate with fractions, using the building blocks of arithmetic.

ADDING FRACTIONS BY CROSS-MULTIPLICATION

Before you can add two fractions, you first have to change them into *like fractions*. And what might that mean? Simply this: Two fractions are like fractions if their bottom numbers are the same.

Definition
Two fractions are like fractions when their bottom numbers are the same.

You have $\frac{1}{3}$ quart of milk in one container and $\frac{1}{2}$ quart in another. If you combine the two portions, will you have enough milk for a recipe calling for one quart of milk? The fractions $\frac{1}{3}$ and $\frac{1}{2}$ are not like fractions, because 3 and 2 are different. However, you can change these two fractions into like fractions. The simple method is to multiply the 3 in the first fraction by 2, and the 2 in the second fraction by 3. This is called *cross-multiplying*.

Definition
To cross-multiply is to multiply both the top number and bottom number in each of two fractions by the bottom number of the other fraction. Both fractions then have the same bottom number.

In both fractions of our example we end up with the bottom number 6. Remember that when you multiply the bottom number by a certain number, you must also multiply the top number by the same number, or you'll change the value of the fraction. Multiply both the top and the bottom of the first fraction by 2.

$$\frac{1}{3} = \frac{1 \times 2}{3 \times 2} = \frac{2}{6}$$

So ²⁄₆ has the same value as ¹⁄₃. Now change ¹⁄₂ into a like fraction by multiplying the top and bottom by 3.

$$\frac{1}{2} = \frac{1 \times 3}{2 \times 3} = \frac{3}{6}$$

Here, ³⁄₆ has the same value as ¹⁄₂. You now have two like fractions— that is, they have equal bottom numbers. Finish the problem by adding the two top numbers and writing the result over the shared bottom number, 6.

$$\frac{2}{6} + \frac{3}{6} = \frac{2 + 3}{6} = \frac{5}{6}$$

You now have ⁵⁄₆ of a quart of milk, which is less than one whole quart. You'll have to go to the store and get more to have enough for that recipe.

Rule

To add two fractions, change them into like fractions (fractions that have the same bottom number), add the top numbers, and write the result over the shared bottom number.

Suppose your niece asks for help with her homework. She needs to find ³⁄₇ + ⁴⁄₅. You can help her. First change the two fractions into like fractions by multiplying each fraction by the bottom number of the other fraction.

Multiply the 3 and 7 of ³⁄₇ by 5.

$$\frac{3}{7} = \frac{3 \times 5}{7 \times 5} = \frac{15}{35}$$

Next multiply the 4 and 5 of ⅘ by 7.

$$\frac{4}{5} = \frac{4 \times 7}{5 \times 7} = \frac{28}{35}$$

Notice that the two fractions now have the same bottom number. Now add the top numbers and write the sum above 35.

$$\frac{15}{35} + \frac{28}{35} = \frac{15 + 28}{35} = \frac{43}{35}$$

The answer to this problem is ⁴³⁄₃₅. Your niece will love you and will get an *A* on her homework. Try the sample problems.

PROBLEMS

Add the following pairs of fractions. Use cross-multiplication.

1. ⅓ + ½ = ?

2. ³⁄₇ + ⁹⁄₁₁ = ?

3. ⁴¹⁄₃ + ¹⁶⁄₅ = ?

4. ¼ + ⅐ = ?

5. ¹¹⁄₇ + ¹⁄₁₆ = ?

6. ⅓ + ⅖ = ?

7. ¹³⁄₂₁ + ¹¹⁄₅₅

8. ³⁄₅₁ + ¹¹⁄₉₁ = ?

9. ³⁰¹⁄₂ + ¹⁴⁄₉₄ = ?

10. ⅕ + ⁶⁄₉ = ?

ANSWERS

1. $\frac{1}{3} + \frac{1}{2} = \frac{1 \times 2}{3 \times 2} + \frac{1 \times 3}{2 \times 3} = \frac{2}{6} + \frac{3}{6} = \frac{(2+3)}{6} = \frac{5}{6}$

2. $\frac{3}{7} + \frac{9}{11} = \frac{3 \times 11}{7 \times 11} + \frac{9 \times 7}{11 \times 7} = \frac{33}{77} + \frac{63}{77} = \frac{(33+63)}{77} = \frac{96}{77}$

3. $\frac{41}{3} + \frac{16}{5} = \frac{41 \times 5}{3 \times 5} + \frac{16 \times 3}{5 \times 3} = \frac{205}{15} + \frac{48}{15} = \frac{(205+48)}{15} = \frac{253}{15}$

4. $\frac{1}{4} + \frac{1}{7} = \frac{1 \times 7}{4 \times 7} + \frac{1 \times 4}{7 \times 4} = \frac{7}{28} + \frac{4}{28} = \frac{(7+4)}{28} = \frac{11}{28}$

5. $\frac{11}{7} + \frac{1}{16} = \frac{11 \times 16}{7 \times 16} + \frac{1 \times 7}{16 \times 7} = \frac{176}{112} + \frac{7}{112} = \frac{(176+7)}{112} = \frac{183}{112}$

6. $\frac{1}{3} + \frac{2}{5} = \frac{1 \times 5}{3 \times 5} + \frac{2 \times 3}{5 \times 3} = \frac{5}{15} + \frac{6}{15} = \frac{(5+6)}{15} = \frac{11}{15}$

7. $\frac{13}{21} + \frac{11}{55} = \frac{13 \times 55}{21 \times 55} + \frac{11 \times 21}{55 \times 21} = \frac{715}{1,155} + \frac{231}{1,155} = \frac{(715+231)}{1,155} = \frac{946}{1,155}$

8. $\frac{3}{51} + \frac{11}{91} = \frac{3\times91}{51\times91} + \frac{11\times51}{91\times51} = \frac{273}{4,641} + \frac{561}{4,641} = \frac{(273+561)}{4,641} = \frac{834}{4,641}$

9. $\frac{301}{2} + \frac{14}{94} = \frac{301\times94}{2\times94} + \frac{14\times2}{94\times2} = \frac{28,294}{188} + \frac{28}{188} = \frac{(28,294+28)}{188} = \frac{28,322}{188}$

10. $\frac{1}{5} + \frac{0}{9} = \frac{1}{5} + 0 = \frac{1}{5}$

ADDING FRACTIONS BY SIMPLIFICATION

When you cross-multiply to change two fractions into like fractions, you often get very large numbers. A possible shortcut is to factor and simplify the fractions first so the numbers you multiply are not so large. However, this "shortcut" can sometimes result in more work than it saves. Bear this in mind as you move ahead.

You can change the two fractions $\frac{1}{3}$ and $\frac{1}{6}$ into like fractions by using cross-multiplication.

$$\frac{1}{3} = \frac{1 \times 6}{3 \times 6} = \frac{6}{18} \quad \text{and} \quad \frac{1}{6} = \frac{1 \times 3}{6 \times 3} = \frac{3}{18}$$

You now have like fractions, but the resulting top numbers and bottom numbers are greater than they have to be. You could have just multiplied the $\frac{1}{3}$ by 2.

$$\frac{1}{3} = \frac{1 \times 2}{3 \times 2} = \frac{2}{6}$$

Now, $\frac{2}{6}$ and $\frac{1}{6}$ are like fractions, and this conversion process involved fewer steps and smaller numbers.

What you need now is a method to find the smallest bottom number that can be shared by the two fractions. To do this, factor the bottom numbers into their primes. Then cross-multiply only those primes that the two bottom numbers do not have in common. It sounds tricky at first, but it gets easier with practice. Work with the above example. The fraction $\frac{1}{3}$ has 3 as the bottom number and 3 is already in its factored form; 6, the bottom number of the second fraction, can be factored into the two primes, 2 and 3.

Original number	Factored number	Original number		Factored number
$\dfrac{1}{3}$	$=\quad\dfrac{1}{3}$	$\dfrac{1}{6}$	$=$	$\dfrac{1}{2\times3}$

The bottom numbers of the two factored fractions share 3 but not 2. What you must do is manipulate the two fractions so that both bottom numbers have the same primes. In the above case you can do this by multiplying both the top and bottom numbers in the fraction $\frac{1}{3}$ by 2.

By factoring you are finding the lowest common bottom number shared by the two fractions. Consider adding the two fractions $\frac{3}{70}$ and $\frac{2}{42}$ by cross-multiplying:

$$\frac{3}{70}+\frac{2}{42}=\frac{3\times42}{70\times42}+\frac{2\times70}{42\times70}=\frac{126}{2,940}+\frac{140}{2,940}=\frac{266}{2,940}$$

Cross-multiplication has resulted in the correct answer, but the answer fraction's bottom number is rather large. When using simplification, we first factor the two bottom numbers.

$$\frac{3}{70}+\frac{2}{42}=\frac{3}{2\times5\times7}+\frac{2}{2\times3\times7}$$

You can see that the only primes not shared by the bottom numbers are 5 and 3. The numbers 2 and 7 are in both bottom numbers. So, adjust the first fraction ($\frac{3}{70}$) by multiplying the top and bottom by 3 (because 3 is absent from its bottom number.) Change the second fraction by multiplying the top and bottom by 5 (because 5 is absent from its bottom number.)

$$\frac{3}{2\times5\times7}+\frac{2}{2\times3\times7}=\frac{3\times3}{2\times5\times7\times3}+\frac{2\times5}{2\times3\times7\times5}$$

The two bottom numbers on the right side of the equation now have the same primes. Multiply the primes, then add the top numbers of the fractions.

$$\frac{3\times3}{2\times5\times7\times3}+\frac{2\times5}{2\times3\times7\times5}=\frac{9}{210}+\frac{10}{210}=\frac{19}{210}$$

Does the answer from cross-multiplication, $^{266}\!/_{2,940}$, have the same value as $^{19}\!/_{210}$, the answer from simplification? Yes. You can see this by factoring out and canceling 14 from the first answer.

$$\frac{266}{2,940} = \frac{19 \times 14}{210 \times 14} = \frac{19}{210}$$

Rule

Adding by simplification. First factor the top and bottom of both fractions, and simplify. Next cross-multiply by only those numbers not shared by the two bottom numbers. Then add the top numbers and write the sum over the common bottom number.

Your niece, who is advancing rapidly in math, has more homework. She wants to add $^{30}\!/_{14} + {}^{8}\!/_{21}$. You can help her out. First factor and simplify both fractions.

$$\frac{30}{14} + \frac{8}{21} = \frac{5 \times 3 \times 2}{2 \times 7} + \frac{2 \times 2 \times 2}{3 \times 7}$$

The fraction $^{(5 \times 3 \times 2)}\!/_{(2 \times 7)}$ has 2 in both the top and bottom numbers. So you can cancel the 2s.

$$\frac{30}{14} + \frac{8}{21} = \frac{5 \times 3 \times \cancel{2}}{\cancel{2} \times 7} + \frac{2 \times 2 \times 2}{3 \times 7} = \frac{5 \times 3}{7} + \frac{2 \times 2 \times 2}{3 \times 7}$$

You now have both fractions in their lowest terms. You can see by inspection that the only whole number that is not in both bottom numbers is 3. You have only to change $^{(5 \times 3)}\!/_{7}$ by multiplying both the top and bottom by 3.

$$\frac{30}{14} + \frac{8}{21} = \frac{5 \times 3}{7} + \frac{2 \times 2 \times 2}{3 \times 7} = \frac{5 \times 3 \times 3}{7 \times 3} + \frac{2 \times 2 \times 2}{3 \times 7}$$

You can now add.

$$\frac{5 \times 3 \times 3}{7 \times 3} + \frac{2 \times 2 \times 2}{3 \times 7} = \frac{45}{21} + \frac{8}{21} = \frac{(45 + 8)}{21} = \frac{53}{21}$$

The answer is $^{53}/_{21}$, and your niece thinks you're a genius. Try some problems.

PROBLEMS

Add each pair of fractions by using simplification.

1. $\frac{1}{2} + \frac{5}{6}$

2. $\frac{3}{20} + \frac{5}{6}$

3. $\frac{1}{30} + \frac{3}{70}$

4. $\frac{3}{22} + \frac{7}{55}$

5. $\frac{2}{60} + \frac{11}{105}$

6. $\frac{5}{12} + \frac{3}{20}$

7. $\frac{3}{34} + \frac{5}{51}$

8. $\frac{6}{36} + \frac{35}{30}$

ANSWERS

1. $\frac{1}{2} + \frac{5}{6} = \frac{1 \times 3}{2 \times 3} + \frac{5}{2 \times 3} = \frac{3}{6} + \frac{5}{6} = \frac{(3+5)}{6} = \frac{8}{6}$

2. $\frac{3}{20} + \frac{5}{6} = \frac{3}{5 \times 2 \times 2} + \frac{5}{2 \times 3} = \frac{3 \times 3}{5 \times 2 \times 2 \times 3} + \frac{5 \times (2 \times 5)}{2 \times 3 \times (2 \times 5)} = \frac{9}{60} + \frac{50}{60} = \frac{(9+50)}{60} = \frac{59}{60}$

3. $\frac{1}{30} + \frac{3}{70} = \frac{1}{5 \times 2 \times 3} + \frac{3}{5 \times 2 \times 7} = \frac{1 \times 7}{5 \times 2 \times 3 \times 7} + \frac{3 \times 3}{5 \times 2 \times 7 \times 3} = \frac{7}{210} + \frac{9}{210} = \frac{(7+9)}{210} = \frac{16}{210}$

4. $\frac{3}{22} + \frac{7}{55} = \frac{3}{2 \times 11} + \frac{7}{5 \times 11} = \frac{3 \times 5}{2 \times 11 \times 5} + \frac{7 \times 2}{5 \times 11 \times 2} = \frac{15}{110} + \frac{14}{110} = \frac{(15+14)}{110} = \frac{29}{110}$

5. $\frac{2}{60} + \frac{11}{105} = \frac{2}{2 \times 2 \times 3 \times 5} + \frac{11}{3 \times 5 \times 7} = \frac{2}{\not{2} \times 2 \times 3 \times 5} + \frac{11}{3 \times 5 \times 7} = \frac{1}{2 \times 3 \times 5} + \frac{11}{3 \times 5 \times 7} = \frac{1 \times 7}{2 \times 3 \times 5 \times 7} + \frac{11 \times 2}{3 \times 5 \times 7 \times 2}$

$= \frac{7}{210} + \frac{22}{210} = \frac{29}{210}$

6. $\frac{5}{12} + \frac{3}{20} = \frac{5}{2 \times 2 \times 3} + \frac{3}{2 \times 2 \times 5} = \frac{5 \times 5}{2 \times 2 \times 3 \times 5} + \frac{3 \times 3}{2 \times 2 \times 5 \times 3} = \frac{25}{60} + \frac{9}{60} = \frac{34}{60}$

7. $\frac{3}{34} + \frac{5}{51} = \frac{3}{2 \times 17} + \frac{5}{3 \times 17} = \frac{3 \times 3}{2 \times 17 \times 3} + \frac{5 \times 2}{3 \times 17 \times 2} = \frac{9}{102} + \frac{10}{102} = \frac{19}{102}$

8. $\frac{6}{36} + \frac{35}{30} = \frac{2 \times 3}{2 \times 2 \times 3 \times 3} + \frac{5 \times 7}{2 \times 3 \times 5} = \frac{\not{2} \times \not{3}}{2 \times \not{2} \times \not{3} \times 3} + \frac{\not{5} \times 7}{2 \times 3 \times \not{5}} = \frac{1}{2 \times 3} + \frac{7}{2 \times 3} = \frac{1}{6} + \frac{7}{6} = \frac{8}{6}$

If you missed more than four of the sample problems, review the sections on cross-multiplication and simplification carefully and try the problems again. Remember to feel free to use a calculator to be sure that your computations are correct. With practice, you'll find both methods easy to use. After adding two fractions, you will often want to reduce the answer to its simplest form by canceling common primes from both the top and bottom numbers.

SUBTRACTING FRACTIONS

To subtract fractions, simply carry out the same procedure as with adding fractions, but where you would add the two top numbers, subtract instead. If your gas gauge shows that you have ½ tank of gas, and you know you need ⅓ tank to drive to the lake and back, how much gas will you have left at the end of your trip? Begin with ½ − ⅓, and use cross-multiplication. Cross-multiplying gives you ³⁄₆ − ²⁄₆. Now subtract 2 from 3 and write the difference over 6.

$$\frac{1}{2} - \frac{1}{3} = \frac{3}{6} - \frac{2}{6} = \frac{3-2}{6} = \frac{1}{6}$$

The answer is ⅙ of a tank. What if you have only ⅓ tank, and your trip requires ½ tank? You subtract ½ from ⅓. Again, change ½ and ⅓ into like fractions.

$$\frac{1}{3} - \frac{1}{2} = \frac{2}{6} - \frac{3}{6} = \frac{2-3}{6} = \frac{(-1)}{6}$$

The answer this time is a negative 1 over a positive 6. But you know that when a fraction's top number and bottom number have opposite signs, the resulting fraction is negative.

$$\frac{(-1)}{6} = -\frac{1}{6}$$

So the answer to ⅓ − ½ is negative ⅙. This had better not be your gas gauge, or you'll have been stranded on the highway for some time now. Look at one more subtraction problem, ²²⁄₁₃ − ⁹⁄₂. First cross-multiply to get like fractions.

$$\frac{22}{13} = \frac{22 \times 2}{13 \times 2} = \frac{44}{26} \quad \text{and} \quad \frac{9}{2} = \frac{9 \times 13}{2 \times 13} = \frac{117}{26}$$

You now have the like fractions ⁴⁴⁄₂₆ and ¹¹⁷⁄₂₆. Subtract.

$$\frac{44}{26} - \frac{117}{26} = \frac{44 - 117}{26} = \frac{-73}{26} = -\frac{73}{26}$$

The answer is negative $\frac{73}{26}$. Once you have mastered the addition of fractions, the subtraction of fractions should be easy.

Rule

To add or subtract fractions, change them into like fractions (fractions that have the same bottom number), do the addition or subtraction with the top numbers, and write the result over the shared bottom number.

Try the sample problems.

PROBLEMS

Solve each subtraction problem by using either cross-multiplication or simplification.

1. $\frac{1}{2} - \frac{1}{4}$

2. $\frac{1}{2} - \frac{1}{5}$

3. $\frac{1}{6} - \frac{1}{5}$

4. $\frac{1}{3} - \frac{1}{11}$

5. $\frac{3}{13} - \frac{5}{9}$

6. $\frac{143}{9} - \frac{1}{2}$

7. $\frac{13}{2} - \frac{0}{5}$

8. $\frac{0}{3} - \frac{1}{4}$

ANSWERS

1. $\frac{1}{2} - \frac{1}{4} = \frac{1}{2} - \frac{1}{2\times 2} = \frac{1\times 2}{2\times 2} - \frac{1}{4} = \frac{2}{4} - \frac{1}{4} = \frac{(2-1)}{4} = \frac{1}{4}$

2. $\frac{1}{2} - \frac{1}{5} = \frac{1\times 5}{2\times 5} - \frac{1\times 2}{5\times 2} = \frac{5}{10} - \frac{2}{10} = \frac{3}{10}$

3. $\frac{1}{6} - \frac{1}{5} = \frac{1\times 5}{6\times 5} - \frac{1\times 6}{5\times 6} = \frac{5}{30} - \frac{6}{30} = -\frac{1}{30}$

4. $\frac{1}{3} - \frac{1}{11} = \frac{1\times 11}{3\times 11} - \frac{1\times 3}{11\times 3} = \frac{11}{33} - \frac{3}{33} = \frac{8}{33}$

5. $\frac{3}{13} - \frac{5}{9} = \frac{3\times 9}{13\times 9} - \frac{5\times 13}{9\times 13} = \frac{27}{117} - \frac{65}{117} = -\frac{38}{117}$

6. $\frac{143}{9} - \frac{1}{2} = \frac{143\times 2}{9\times 2} - \frac{1\times 9}{2\times 9} = \frac{286}{18} - \frac{9}{18} = \frac{277}{18}$

7. $\frac{13}{2} - \frac{0}{5} = \frac{13}{2} - 0 = \frac{13}{2}$

8. $\frac{0}{3} - \frac{1}{4} = 0 - (^{+}\frac{1}{4}) = 0 + (^{-}\frac{1}{4}) = -\frac{1}{4}$

Notice that problems 7 and 8 required neither cross-multiplication nor simplification. You needed only your knowledge of rules about 0 and subtraction.

MULTIPLYING FRACTIONS

Multiplying fractions is actually easier than adding or subtracting. When multiplying two fractions, simply multiply the two top numbers, then multiply the two bottom numbers. You don't have to worry about like fractions, cross-multiplication, or simplification.

Rule
To multiply two fractions, multiply the top numbers, multiply the bottom numbers, and write the product of the top numbers over the product of the bottom numbers.

You and a friend buy a bag of popcorn and split it evenly in half. Your brother comes along and takes ⅓ of your half. How much of the original bag did he take?

$$\frac{1}{2} \times \frac{1}{3} = \frac{1 \times 1}{2 \times 3} = \frac{1}{6}$$

Here comes your niece again, announcing that (⅔) × (5/7) equals 10/21. Is she correct?

Multiply 2 × 5 and write the answer on top. Then multiply 3 × 7 and write that answer on the bottom.

$$\frac{2}{3} \times \frac{5}{7} = \frac{2 \times 5}{3 \times 7} = \frac{10}{21}$$

Your niece was absolutely correct. Try one more.

$$\frac{23}{6} \times \frac{3}{4} = \frac{23 \times 3}{6 \times 4} = \frac{69}{24}$$

You're ready to try the sample problems. Remember to use a calculator when you think you need to.

PROBLEMS

Multiply each pair of fractions.

1. $\frac{3}{4} \times \frac{6}{7}$

2. $\frac{1}{3} \times \frac{7}{9}$

3. $\frac{11}{2} \times \frac{9}{5}$

4. $\frac{22}{9} \times \frac{43}{2}$

5. $\frac{142}{13} \times \frac{1}{2}$

6. $\frac{90}{13} \times \frac{101}{203}$

7. $\frac{1}{103} \times \frac{45}{11}$

8. $\frac{17}{3} \times \frac{1,007}{16}$

9. $\frac{0}{19} \times \frac{123}{7}$

10. $\frac{1,754}{15} \times \frac{13}{2,811}$

ANSWERS

1. $\frac{3}{4} \times \frac{6}{7} = \frac{3 \times 6}{4 \times 7} = \frac{18}{28}$

2. $\frac{1}{3} \times \frac{7}{9} = \frac{1 \times 7}{3 \times 9} = \frac{7}{27}$

3. $\frac{11}{2} \times \frac{9}{5} = \frac{11 \times 9}{2 \times 5} = \frac{99}{10}$

4. $\frac{22}{9} \times \frac{43}{2} = \frac{22 \times 43}{9 \times 2} = \frac{946}{18}$

5. $\frac{142}{13} \times \frac{1}{2} = \frac{142 \times 1}{13 \times 2} = \frac{142}{26}$

6. $\frac{90}{13} \times \frac{101}{203} = \frac{90 \times 101}{13 \times 203} = \frac{9,090}{2,639}$

7. $\frac{1}{103} \times \frac{45}{11} = \frac{1 \times 45}{103 \times 11} = \frac{45}{1,133}$

8. $\frac{17}{3} \times \frac{1,007}{16} = \frac{17 \times 1,007}{3 \times 16} = \frac{17,119}{48}$

9. $\frac{0}{19} \times \frac{123}{7} = \frac{0 \times 123}{19 \times 7} = \frac{0}{133} = 0$

10. $\frac{1,754}{15} \times \frac{13}{2,811} = \frac{1,754 \times 13}{15 \times 2,811} = \frac{22,802}{42,165}$

Now that you're a master at multiplying fractions, maybe you'd like to jump one step ahead of your niece by quickly conquering division of fractions.

DIVIDING FRACTIONS

Division of fractions is as simple as multiplication, but it requires one extra step. You change the division problem into a multiplication problem by inverting the fraction by which you are dividing. To invert a fraction, simply exchange the top and bottom numbers.

Rule
To divide with fractions, invert the fraction by which you are dividing, then multiply.

Divide ¾ by 5⁄7.

$$\frac{3}{4} \div \frac{5}{7} = ?$$

To do this division, invert the fraction 5⁄7 and multiply.

$$\frac{3}{4} \times \frac{7}{5} = \frac{3 \times 7}{4 \times 5} = \frac{21}{20}$$

So 21⁄20 is the result of dividing ¾ by 5⁄7. You could show your niece how to divide 11⁄32 by 17⁄9.

$$\frac{11}{32} \div \frac{17}{9} = ?$$

Invert the number you are dividing by, 17⁄9, and multiply.

$$\frac{11}{32} \times \frac{9}{17} = \frac{11 \times 9}{32 \times 17} = \frac{99}{544}$$

The answer is 99⁄544. As you can see, the division of fractions is simple. Try some problems. (Or, why not have a race with your niece?)

PROBLEMS

Solve each division problem.

1. ½ ÷ ⅔   **4.** ⅗ ÷ ¼

2. 9⁄2 ÷ 3⁄1   **5.** 15⁄4 ÷ 11⁄9

3. 2⁄51 ÷ 1⁄33   **6.** 113⁄9 ÷ 431⁄7

ANSWERS

1. $\frac{1}{2} \div \frac{2}{3} = \frac{1}{2} \times \frac{3}{2} = \frac{1 \times 3}{2 \times 2} = \frac{3}{4}$ **2.** $\frac{9}{2} \div \frac{3}{1} = \frac{9}{2} \times \frac{1}{3} = \frac{9 \times 1}{2 \times 3} = \frac{9}{6}$

3. $\frac{2}{51} \div \frac{1}{33} = \frac{2}{51} \times \frac{33}{1} = \frac{2 \times 33}{51 \times 1} = \frac{66}{51}$

4. $\frac{3}{5} \div \frac{1}{4} = \frac{3}{5} \times \frac{4}{1} = \frac{3 \times 4}{5 \times 1} = \frac{12}{5}$

5. $\frac{15}{4} \div \frac{11}{9} = \frac{15}{4} \times \frac{9}{11} = \frac{15 \times 9}{4 \times 11} = \frac{135}{44}$

6. $\frac{113}{9} \div \frac{431}{7} = \frac{113}{9} \times \frac{7}{431} = \frac{113 \times 7}{9 \times 431} = \frac{791}{3,879}$

MIXED NUMBERS AND PURE FRACTIONS

Sometimes fractions are written as mixed numbers such as $3\frac{1}{2}$. When manipulating fractions, it is helpful to change this mixed form into a pure fraction. Remember that $3\frac{1}{2}$ can be considered as the number 3 plus the number $\frac{1}{2}$.

$$3\frac{1}{2} = 3 + \frac{1}{2}$$

Write the 3 as a fraction, or $\frac{3}{1}$.

$$3\frac{1}{2} = \frac{3}{1} + \frac{1}{2}$$

Now add the fractions.

$$\frac{3}{1} + \frac{1}{2} = \frac{3 \times 2}{1 \times 2} + \frac{1}{2} = \frac{6}{2} + \frac{1}{2} = \frac{7}{2}$$

An easier way to do this is to multiply the whole number by the bottom number in the fraction; then add that to the top number. Put the sum in a new fraction over the same bottom number as the one in the old fraction. For $3\frac{1}{2}$, multiply the whole number 3 by the 2. This gives you 6. Add that to the 1 in $\frac{1}{2}$. The sum is 7. Put the 7 over the old bottom number, 2. Your answer is $\frac{7}{2}$.

PROBLEMS

Change each mixed number to a pure fraction.

1. $2\frac{1}{3}$

2. $1\frac{1}{99}$

3. $8\frac{1}{3}$

4. $9\frac{1}{4}$

5. $16\frac{1}{2}$

ANSWERS

1. $2\frac{1}{3} = \frac{2}{1} + \frac{1}{3} = \frac{2\times3}{1\times3} + \frac{1}{3} = \frac{6}{3} + \frac{1}{3} = \frac{7}{3}$ or $2 \times 3 = 6; 6 + 1 = 7; \frac{7}{3}$

2. $1\frac{1}{99} = \frac{1}{1} + \frac{1}{99} = \frac{1\times99}{1\times99} + \frac{1}{99} = \frac{99}{99} + \frac{1}{99} = \frac{100}{99}$ or $1 \times 99 = 99; 99 + 1 = 100; 100/99$

3. $8\frac{1}{3} = \frac{8}{1} + \frac{1}{3} = \frac{8\times3}{1\times3} + \frac{1}{3} = \frac{24}{3} + \frac{1}{3} = \frac{25}{3}$ or $8 \times 3 = 24; 24 + 1 = 25; \frac{25}{3}$

4. $9\frac{1}{4} = \frac{9}{1} + \frac{1}{4} = \frac{9\times4}{1\times4} + \frac{1}{4} = \frac{36}{4} + \frac{1}{4} = \frac{37}{4}$ or $9 \times 4 = 36; 36 + 1 = 37; \frac{37}{4}$

5. $16\frac{1}{2} = \frac{16}{1} + \frac{1}{2} = \frac{16\times2}{1\times2} + \frac{1}{2} = \frac{32}{2} + \frac{1}{2} = \frac{33}{2}$ or $16 \times 2 = 32; 32 + 1 = 33; \frac{33}{2}$

WHAT HAVE YOU LEARNED?

You have learned how to add, subtract, multiply, and divide fractions.

1. Definition: Two fractions are *like fractions* when their bottom numbers are the same (e.g. $\frac{2}{6}$ and $\frac{3}{6}$).
2. Definition: To *cross-multiply* is to multiply both the top and bottom number of two fractions by the bottom number of the other fraction.
3. Rule: Adding by simplification. First factor the top and bottom of both fractions, and simplify. Next cross-multiply by only those numbers not shared by the two bottom numbers. Then add the top numbers, and write the sum over the shared bottom number.

$$\frac{30}{14} + \frac{8}{21} = \frac{5 \times 3 \times \cancel{2}}{\cancel{2} \times 7} + \frac{2 \times 2 \times 2}{3 \times 7} = \frac{5 \times 3}{7} + \frac{2 \times 2 \times 2}{3 \times 7}$$

$$= \frac{5 \times 3 \times 3}{7 \times 3} + \frac{2 \times 2 \times 2}{3 \times 7} = \frac{45}{21} + \frac{8}{21} = \frac{(45 + 8)}{21} = \frac{53}{21}$$

4. Rule: To add or subtract fractions, change them into like fractions (fractions that have the same bottom number), do the addition or subtraction with the top numbers, and write the result over the common bottom number.

$$\frac{1}{2} - \frac{1}{3} = \frac{3}{6} - \frac{2}{6} = \frac{3 - 2}{6} = \frac{1}{6}$$

5. Rule: To multiply two fractions, multiply the top numbers, multiply the bottom numbers, and write the product of the top numbers over the product of the bottom numbers.

$$\frac{2}{3} \times \frac{5}{7} = \frac{2 \times 5}{3 \times 7} = \frac{10}{21}$$

6. Rule: To divide with fractions, invert the fraction by which you are dividing, then multiply.

$$\frac{3}{4} \div \frac{5}{7} = \frac{3}{4} \times \frac{7}{5} = \frac{3 \times 7}{4 \times 5} = \frac{21}{20}$$

Now that you're an expert at manipulating fractions, the next chapter will seem like a day at the beach. On to the decimal dunes!

CHAPTER 11

DECIMALS

Welcome to the four basic operations of arithmetic with decimals. Decimals are important because they are the foundation of our money system. As you progress through this chapter, you will see why decimals have replaced fractions in many applications. Doing computations with decimals is easier than with fractions, and calculators show results in decimals.

ADDING AND SUBTRACTING DECIMALS

Adding or subtracting decimals is extremely easy. To add two decimals, simply set them up as a long-addition problem and carry out the long addition. The one thing to remember is that you must always stack the decimal points directly over one another.

Suppose you have 107.991 shares of stock, and your dividend for the month is 3.4298 shares. How many shares do you have now? Add 3.4298 and 107.991.

$$
\begin{array}{r}
3.4\,2\,9\,8 \\
+\ 1\ 0\ 7.9\,9\,1 \\
\hline
\end{array}
$$

Align the decimal points over one another. Now add the 8 at the far right of the top number into the blank below it. When you have such a blank, treat it as a 0.

$$
\begin{array}{r}
3.4\,2\,9\,8 \\
+\ 1\ 0\ 7.9\,9\,1\,0 \\
\hline
8
\end{array}
$$

Now move one place to the left and add 9 + 1 to get 10. Write the 0 and carry the 1.

$$
\begin{array}{r}
(1) \\
3\,.\,4\ \ 2\ \ 9\ \ 8 \\
+\ 1\ 0\ 7\,.\,9\ \ 9\ \ 1\ \ 0 \\
\hline
0\ \ 8
\end{array}
$$

In the third column from the right, add 2 + 9 plus the carried 1 to get 12. Write the 2 and carry the 1.

$$
\begin{array}{r}
(1)(1) \\
3\,.\,4\ \ 2\ \ 9\ \ 8 \\
+\ 1\ 0\ 7\,.\,9\ \ 9\ \ 1\ \ 0 \\
\hline
2\ \ 0\ \ 8
\end{array}
$$

Repeat the same steps in the next column to the left.

$$
\begin{array}{r}
(1)(1)(1) \\
3\,.\,4\ \ 2\ \ 9\ \ 8 \\
+\ 1\ 0\ 7\,.\,9\ \ 9\ \ 1\ \ 0 \\
\hline
4\ \ 2\ \ 0\ \ 8
\end{array}
$$

Now write a decimal point in the answer, to the left of the 4, directly below the other decimal points.

$$
\begin{array}{r}
(1)(1)(1) \\
3\,.\,4\ \ 2\ \ 9\ \ 8 \\
+\ 1\ 0\ 7\,.\,9\ \ 9\ \ 1\ \ 0 \\
\hline
.\,4\ \ 2\ \ 0\ \ 8
\end{array}
$$

Continue with the addition.

$$
\begin{array}{r}
(1)(1)(1)(1) \\
3\,.\,4\ \ 2\ \ 9\ \ 8 \\
+\ 1\ 0\ 7\,.\,9\ \ 9\ \ 1\ \ 0 \\
\hline
1\ \ 1\ 1\,.\,4\ \ 2\ \ 0\ \ 8
\end{array}
$$

You have 111.4208 shares of stock. The only difference between this procedure and long addition of whole numbers is that you align the decimal points and include a decimal point in the answer.

Subtracting is just as simple. Set up the problem as in long subtraction, but line up the decimal points and include a decimal point in the answer. Suppose another stock, of which you owned 107.54 shares, suffered a 34.901-share loss. What would your new holdings be? Subtract 34.901 from 107.54.

$$
\begin{array}{r}
1\ 0\ 7.5\ 4 \\
-\ \ \ 3\ 4.9\ 0\ 1 \\
\hline
\end{array}
$$

The top decimal has two digits to the right of the decimal point, but the bottom decimal has three. Fill the empty space by writing a zero above the 1.

$$
\begin{array}{r}
1\ 0\ 7.5\ 4\ 0 \\
-\ \ \ 3\ 4.9\ 0\ 1 \\
\hline
\end{array}
$$

To subtract the 1 at the far right from the 0, borrow from the 4 in the next place to the left. (You can review this procedure by looking back at Chapter 5.)

$$
\begin{array}{r}
3 \\
1\ 0\ 7.5\ \cancel{4}\,10 \\
-\ \ \ 3\ 4.9\ 0\ 1 \\
\hline
\end{array}
$$

Now you can subtract 1 from 10 to get 9.

$$
\begin{array}{r}
3 \\
1\ 0\ 7.5\ \cancel{4}\,10 \\
-\ \ \ 3\ 4.9\ 0\ 1 \\
\hline
9 \\
\end{array}
$$

Proceed by subtracting 0 from 3 to get 3.

```
                     3
            1 0 7. 5 4̸ 10
          −   3 4. 9 0  1
          ─────────────────
                       3  9
```

To subtract 9 from 5, borrow again.

```
                 6   3
            1 0 7̸.15 4̸ 10
          −   3 4. 9 0  1
          ─────────────────
                       3  9
```

Subtract 9 from 15 to get 6.

```
                 6   3
            1 0 7̸.15 4̸ 10
          −   3 4. 9 0  1
          ─────────────────
                     6 3  9
```

You have reached the decimal point, so write it in the answer.

```
                 6   3
            1 0 7̸.15 4̸ 10
          −   3 4. 9 0  1
          ─────────────────
                   . 6 3  9
```

Continue to subtract.

```
             0   6   3
           1̸ 10 7̸.15 4̸ 10
          −   3 4. 9 0  1
          ─────────────────
               7 2. 6 3  9
```

In the last step, you are subtracting a blank from a 0, which gives you nothing to write at the left of the 7 in the answer. Your new stock holding is 72.639 shares.

After considering the following rule, try the sample problems.

Rule

To add or subtract decimals, stack them with their decimal points lined up and apply long addition or long subtraction, writing a decimal point in the appropriate position in the answer.

PROBLEMS

1. 2.3 + 1.0 = ?

2. 1.1 − 0.1 = ?

3. 3.4 − 4.4 = ?

4. 3.66 + 99 = ?

5. 0.117 + 1.4 = ?

6. 2.309 + 14.01 = ?

7. 1.23 + 0.00011 = ?

8. 1.0117 − 3.9 = ?

9. 67.00917 − 3.59732 = ?

10. 1 − 0.00001 = ?

ANSWERS

1.
```
   2.3
 + 1.0
 ─────
   3.3
```

2.
```
   1.1
 - 0.1
 ─────
   1.0
```

3.
```
   4.4
 - 3.4
 ─────
 (-1.0)
```

4.   (1)
```
    3.6 6
 +  9 9.0 0
 ─────────
 1 0 2.6 6
```

5.
```
   0.1 1 7
 + 1.4 0 0
 ─────────
   1.5 1 7
```

6.
```
    2.3 0 9
 + 1 4.0 1 0
 ───────────
 1 6.3 1 9
```

7.
```
    1.2 3 0 0 0
 +  0.0 0 0 1 1
 ──────────────
    1.2 3 0 1 1
```

8.
```
      8 9 9 10
   3.9̸ 0̸ 0̸ 0̸
 -  1.0 1 1 7
 ────────────
 (-2.8 8 8 3)
```

9.
```
    6 9 10 8 11
 6 7.0̸ 0̸9̸ 1̸ 7
 -   3.5 9 7 3 2
 ───────────────
 6 3.4 1 1 8 5
```

10.
```
    0 9 9 9 9 10
 1̸.0̸ 0̸ 0̸ 0̸ 0̸
 -  0.0 0 0 0 1
 ──────────────
    0.9 9 9 9 9
```

MULTIPLYING DECIMALS

To multiply decimals, use long multiplication and then include the decimal point. Look at a simple example. Multiply 341 by 0.5. Set up the problem as a long-multiplication problem.

$$\begin{array}{r} 3\ 4\ 1 \\ \times\ \ \ 0.\,5 \\ \hline \end{array}$$

Here you don't need to line up the decimals (the decimal in the top number would be at the right of the 1 if it were written). First multiply the top numbers by the 5 on the bottom.

$$\begin{array}{r} (2) \\ 3\ 4\ 1 \\ \times\ \ \ 0.\,5 \\ \hline 1\ 7\ 0\ 5 \end{array}$$

The far-left digit in the bottom number is a 0, and 0 times any number is 0. Your multiplication is finished. You have the correct digits for the answer, 1705, but you need to determine where to place the decimal point in the answer. Do this by counting all the digits to the right of the decimal points in both numbers being multiplied. You can see that the top number has no digits to the right of its decimal point, because the decimal point is not written. The bottom number has one digit to the right of the decimal point. Therefore, the top number contributes zero places, and the bottom number contributes 1 decimal place. From the far right of 1705, count one place to the left, and write the decimal point. When you add the decimal point, 1705 becomes 170.5, which is the answer.

Suppose your 5.19 shares of Space Technology stock generated $17.05 per share last month. How much money did your shares produce? Multiply $17.05 by 5.19.

$$\begin{array}{r} 1\ 7.0\ 5 \\ \times\ \ \ 5.1\ 9 \\ \hline \end{array}$$

Carry out the long multiplication; ignore the decimal points for the moment.

```
          (3) (2)
          (6) (4)
          1 7.0 5
      ×        5.1 9
      ─────────────
      (1)
      1 5 3 4 5
      1 7 0 5
    8 5 2 5
    ─────────────
    8 8 4 8 9 5
```

Carefully review the above problem; the multiplication has been done according to the long-multiplication procedure. Now you have to find out where to place the decimal point. The top number has two digits to the right of the decimal point, and the bottom number also has two digits to the right of the decimal point. The sum of 2 and 2 is 4. Count 4 digits from the right end of the answer and write the decimal point. The 884895 in the above problem becomes the answer 88.4895. Our stock produced $88.4895 last month—quite a profit.

After reviewing the following rule, try the sample problems.

Rule

To multiply decimals, use the long-multiplication procedure. Then count the number of digits to the right of the decimal point in each number you are multiplying, and add them. Beginning at the right end of the answer, move left this number of digits and write the answer's decimal point.

PROBLEMS

1. 4.5 × 2.2

2. 0.11 × 3.2

3. 0.04 × 1.01

4. 12.1 × 1.21

5. 0.015 × 4.32

6. 0.0015 × 3

ANSWERS

1. (1)
(1)
4.5
× 2.2
─────
9 0
+ 9 0
─────
9.9 0

2. 0.1 1
× 3.2
─────
0 2 2
+0 3 3
─────
0.3 5 2

3. 1.0 1
× 0.0 4
─────
4 0 4
0 0 0
+0 0 0
─────
0.0 4 0 4

4. 1 2.1
× 1.2 1
─────
1 2 1
2 4 2
+1 2 1
─────
1 4.6 4 1

5. (2)
(1)
(1)
0.0 1 5
× 4.3 2
─────
0 0 3 0
0 0 4 5
+0 0 6 0
─────
0.0 6 4 8 0

6. (1)
0.0 0 1 5
× 3
─────
0.0 0 4 5

DIVIDING DECIMALS

Division with decimals is simply an extension of the long-division process. Handling the decimal point in division is, however, a little more complicated than it was with multiplication. (Isn't everything about division a little more complicated?) But you'll learn a handy shortcut.

Suppose your daughter needs to divide 45 by 3.2. How would you begin to explain it to her? Set up the problem as usual.

$$3.2 \overline{\smash)4\ 5}$$

Before dividing, you are going to perform a little trick to change this problem into a regular long-division problem. You are going to get rid of the decimal point in the number doing the dividing. Simply multiply both the number being divided and the number doing the dividing by 10. Here's the result.

$$3\,2\,\overline{\smash{\big)}\,4\,5\,0}$$

Why can you do this? Consider our original division problem as a fraction.

$$\frac{45}{3.2}$$

You know from working with fractions that you can multiply both the top number and the bottom number by the same number without changing the value of the fraction. Multiplying the top and bottom of the above fraction by 10 produces the following result.

$$\frac{45 \times 10}{3.2 \times 10} = \frac{450}{32}$$

Is it allowable to change the problem and get rid of the decimal point? Absolutely. What makes the process nice is that you always multiply both numbers in the division problem by the same number, a number consisting of a 1 followed by a number of zeros, that is, you multiply by 10 or 100 or 1000, etc. The number you multiply by has the same number of zeros as the number of digits to the right of the decimal point in the number doing the dividing. If you are dividing by a number with one digit to the right of the decimal point, then you multiply by 10. If it has two numbers right of the decimal point, then you multiply by 100.

In fact, you can take a shortcut. The effect of multiplying both numbers in the division problem by the same multiple of 10 is simply to shift the digits of each number toward the left of the number's decimal point, which results in shifting the decimal point, itself, to the right. Remember: shifting digits *left* causes the decimal point to shift *right*. You want to shift the decimal point in the number doing the dividing enough places to bring it to the extreme right. Then you can drop the decimal point. Next, shift the decimal point of the number being divided toward the right the same number of places, writing zeros at the right, if necessary.

Try another example: .0119 divided by 10.11. Looks messy, doesn't it? But you can make it simple. Set up the problem.

$$1\,0.1\,1\,\overline{\,)\,.0\,1\,1\,9}$$

Now move the digits in the dividing number two places toward the left, until the decimal point is at the extreme right.

$$1\,0\,1\,1.\,\overline{\,)\,.0\,1\,1\,9}$$

In its present position, the decimal point isn't really needed, and you can drop it.

$$1\,0\,1\,1\,\overline{\,)\,.0\,1\,1\,9}$$

Now shift the number being divided two places toward the left, too.

$$1\,0\,1\,1\,\overline{\,)\,0\,1.1\,9}$$

You can now proceed with long division. You need to keep the decimal point in the proper position in the answer. The easy way to do this is to write the answer's decimal point directly above the decimal point that is inside the division housing.

$$1\,0\,1\,1\,\overline{\,)\,0\,1\overset{\textstyle .}{.}1\,9}$$

Now begin to divide. Will 1011 divide into 0, the first digit at the left? No. Move one digit to the right and pick up the 1. Will 1011 divide into 1? No. Moving again to the right, pick up the digit to the right of the decimal point. Will 1011 divide into 11? No. Now that you are dealing with a digit on the right of the answer's decimal point, you want to freeze the decimal point's position in the answer. To do this, write a 0 above the 1.

$$\overset{\textstyle .0}{1\,0\,1\,1\,\overline{\,)\,0\,1.1\,9}}$$

Will 1011 divide into 119? No. Write a 0 above the 9.

```
          . 0 0
        ┌─────────
1 0 1 1 │ 0 1. 1 9
```

Now what? You've run out of digits in the number being divided, and you have only zeros in the answer. But you're not finished yet. You can add zeros at the right end of a decimal without changing the value of the decimal. Write two zeros.

```
          . 0 0
        ┌─────────────
1 0 1 1 │ 0 1. 1 9 0 0
```

Now pick up one of those zeros you've added to get 1190. Will 1011 divide into 1190? Yes! It will divide once but not twice. Write a 1 in the answer and proceed.

```
          . 0 0 1
        ┌─────────────
1 0 1 1 │ 0 1. 1 9 0 0
          1 0 1 1
          ─────────
            1 7 9
```

Bring down the next 0 and tack it onto the 179.

```
          . 0 0 1
        ┌─────────────
1 0 1 1 │ 0 1. 1 9 0 0
          1 0 1 1 x
          ─────────
            1 7 9 0
```

The 1011 will also divide into 1790 once but not twice.

```
          . 0 0 1 1
        ┌─────────────
1 0 1 1 │ 0 1. 1 9 0 0
          1 0 1 1 x
          ───────────
            1 7 9 0
            1 0 1 1
          ───────────
              7 7 9
```

Are you done? That depends on whether you want to be done. Is the answer .0011 accurate enough for your needs? If not, write two more zeros at the right of the number being divided, and continue. Without showing each step, here's the full answer.

```
                   . 0  0  1  1  7  7
      1 0 1 1 | 0 1. 1  9  0  0  0  0
                 1 0  1  1  x  x  x
                 _____
                   1  7  9  0
                   1  0  1  1
                   _____
                      7  7  9  0
                      7  0  7  7
                      _____
                         7  1  3  0
                         7  0  7  7
                         _____
                            5  3
```

You now have .001177 for an answer. If the decimal 0.001177 fulfills your computational needs, you are done. If not, write more zeros, and continue. If the answer is a terminating decimal, the process will eventually result in a remainder of 0 and come to an end. If the answer is an infinite repeating decimal, the process will never end by itself. You must decide where to end it. Before trying the sample problems, review the essentials for dividing decimals, as shown in the following display.

Suppose you have 3.5 ounces of special mosquito-fighting cream to divide among 7.3 people (one of the people is a very small child, hence the .3). Divide 3.5 by 7.3.

(setting up the problem)
$$7.3 \overline{)3.5}$$

(shifting the decimal point to the right)
$$7\,3. \overline{)3.5}$$

(dropping the decimal point)
$$7\,3 \overline{)3.5}$$

hifting to the right under the housing)

$$73 \overline{\smash{)}\ 3\ 5.}$$

lacing the answer's decimal point)

$$73 \overline{\smash{)}\ 3\ 5.}\quad ^{\displaystyle .}$$

vriting a 0 under the housing)

$$73 \overline{\smash{)}\ 3\ 5.0}\quad ^{\displaystyle .}$$

irst long division)

$$
\begin{array}{r}
.4 \\
73 \overline{\smash{)}\ 3\ 5.0} \\
2\ 9\ 2 \\
\hline
5\ 8
\end{array}
$$

vriting two zeros for more accuracy)

$$
\begin{array}{r}
.4 \\
73 \overline{\smash{)}\ 3\ 5.0\ 0\ 0} \\
2\ 9\ 2 \\
\hline
5\ 8
\end{array}
$$

continuing long division)

$$
\begin{array}{r}
.4\ 7\ 9 \\
73 \overline{\smash{)}\ 3\ 5.0\ 0\ 0} \\
2\ 9\ 2\ x\ x \\
\hline
5\ 8\ 0 \\
5\ 1\ 1 \\
\hline
6\ 9\ 0 \\
6\ 5\ 7 \\
\hline
3\ 3
\end{array}
$$

The answer is .479. Because there is a small remainder after the
ist long division, you could continue the division if you wanted more
recision in the answer.
Try some sample problems.

PROBLEMS

Divide.

1. 5 ⟌ 3.4 = ?

2. 1.2 ⟌ .0 1 1 = ?

3. 0.0 7 ⟌ 1 0 1 = ?

4. 6 6.1 ⟌ 5.1 1 = ?

5. 9 9.1 ⟌ 7 7.7 = ?

6. 0.1 0 4 ⟌ 7 3.9 1 = ?

ANSWER

Your answers may be longer or shorter than the printed answers if you carried your answers out different numbers of places.

1.
```
      . 6 8
  5 ⟌ 3.4 0
      3 0 x
      ─────
        4 0
        4 0
      ─────
          0  (remainder)
```

2.
```
                      0.0 0 9 1
  1.2 ⟌ .0 1 1   =   1 2 ⟌ 0.1 1 0 0
                          1 0 8
                          ─────
                            2 0
                            1 2
                          ─────
                              8  (remainder)
```

3.
```
                      1 4 4 2.
  0.0 7 ⟌ 1 0 1   =   7 ⟌ 1 0 1 0 0.
                          7 x x x
                          ───────
                          3 1
                          2 8
                          ─────
                            3 0
                            2 8
                          ─────
                              2 0
                              1 4
                          ─────
                                6  (remainder)
```

```
                          .0 7 7 3
6 6.1 ⌐5.1 1   =    6 6 1 ⌐5 1.1 0 0 0
                          4 6 2 7 x x
                          ─────────
                          4 8 3 0
                          4 6 2 7
                          ─────────
                          2 0 3 0
                          1 9 8 3
                          ─────────
                            4 7  (remainder)


                          .7 8 4
9 9.1 ⌐7 7.7   =    9 9 1 ⌐7 7 7.0 0 0
                          6 9 3 7 x x
                          ─────────
                          8 3 3 0
                          7 9 2 8
                          ─────────
                          4 0 2 0
                          3 9 6 4
                          ─────────
                            5 6  (remainder)


                          7 1 0.6 7
0.1 0 4 ⌐7 3.9 1   =  1 0 4 ⌐7 3 9 1 0.0 0
                          7 2 8 x x x x
                          ─────────
                          1 1 1
                          1 0 4
                          ─────────
                            7 0 0
                            6 2 4
                          ─────────
                            7 6 0
                            7 2 8
                          ─────────
                            3 2  (remainder)
```

WHAT HAVE YOU LEARNED?

Rule: To add or subtract decimals, stack them with their decimal points lined up and apply long addition or long subtraction, writing a decimal point in the appropriate position in the answer.

Rule: To multiply decimals, use long multiplication. Then count the number of digits to the right of the decimal point in each decimal, and add them. Beginning at the right in the answer, count this number of digits to the left and write the answer's decimal point.

To divide decimals, follow Rules 3 through 7.

Rule: Eliminate the decimal point from the number doing the division by shifting its digits to the left of its decimal point.

4. Rule: Shift the digits to the left in the number being divided (insid the division housing). Shift the same number of places as you did the number outside the division housing, writing zeros if necessar

5. Rule: Write the answer's decimal point directly above the decim point that is inside the division housing.

6. Rule: For greater accuracy, write zeros on the right side of the de imal point—at the right end of the number being divided.

7. Rule: Carry out long division, remembering to lock the answe decimal point into position with zeros when necessary.

If you had any trouble with dividing decimals, review the example c pages 110 and 111. Then try some problems again.

Now you're ready to change fractions to decimals.

CHAPTER 12

CHANGING FRACTIONS
INTO DECIMALS

In this chapter, you will convert fractions to decimals and decimals to fractions. You will also be introduced to mixed-form numbers, or numbers made up of combinations of fractions, decimals, and whole numbers.

CHANGING FRACTIONS TO DECIMALS

Because decimals are simply another way of writing fractions, you can move easily from expressing a number in one form to expressing it in another. You will find it helpful to be able to do this, because it can make doing computations much easier.

Rule
To change a fraction to a decimal, divide the fraction's top number by its bottom number.

This is a perfect opportunity to use your calculator. If your fraction is $^3/_{16}$, you simply use long division to divide 16 into 3 and get 0.1875. If the fraction's top number is greater than the bottom number, you'll get a number greater than 1. For example, the decimal for $^{17}/_2$ is 8.5. Whenever the bottom number in the fraction breaks down, or factors, into a combination of 2s and/or 5s, the resulting decimal will be a terminating decimal. However, if the fraction's bottom number contains another prime number besides 2 and 5 when the fraction is in simplest form, the resulting decimal will be an infinite, repeating decimal. In such cases, you need to decide where to stop the division process. Get your feet wet by trying the sample problems.

115

PROBLEMS

Change the following fractions to decimals. When the division process does not end naturally, continue dividing until the answer has at least four digits to the right of the decimal point.

1. ¾   **4.** ¹⁶⁄₂₅

2. ⅜   **5.** ⅓

3. ¹¹⁄₂₀   **6.** ¹⁄₁₇

ANSWERS

1.

```
        . 7 5
   4 | 3.0 0
       2 8 x
       ─────
         2 0
         2 0
       ─────
           0
```

4.

```
          . 6 4
   2 5 | 1 6.0 0
         1 5 0 x
         ───────
           1 0 0
           1 0 0
         ───────
               0
```

2.

```
        .3 7 5
   8 | 3.0 0 0
       2 4 x x
       ───────
         6 0
         5 6
       ───────
           4 0
           4 0
       ───────
             0
```

5.

```
          .3 3 3 3
   3 | 1.0 0.0 0
       9 x x x
       ───────
       1 0
         9
       ───────
         1 0
           9
       ───────
           1 0
             9
       ───────
             1   (remainder)
```

3.

```
          .5 5
   2 0 | 1 1.0 0
         1 0 0 x
         ───────
           1 0 0
           1 0 0
         ───────
               0
```

6.

```
          .0 5 8 8
   1 7 | 1.0 0 0 0
         8 5 x x
         ───────
         1 5 0
         1 3 6
         ───────
           1 4 0
           1 3 6
         ───────
               4   (remainder)
```

CHANGING TERMINATING DECIMALS TO FRACTIONS

Every terminating decimal can be converted to a fraction. First express the decimal as a fraction by writing it over 1.

$$0.35 = \frac{0.35}{1}$$

Remember that dividing or multiplying a number by 1 does not change the value of the number. Now remove the decimal point from the top number of the fraction by multiplying both the top and the bottom of the fraction by a number consisting of a 1 followed by zeros. Sound familiar? That's exactly how you eliminated the decimal point in long division of decimals. For the above example, multiply the top and bottom by 100 because 100 has two zeros—the number of decimal places in the top number.

$$\frac{0.35 \times 100}{1 \times 100} = \frac{35}{100}$$

You now have the fraction $^{35}/_{100}$. But you've probably noticed that it is not in its simplest form. You could reduce it to $^{7}/_{20}$ by factoring both the top and the bottom numbers and then canceling like primes as you learned in Chapter 10. It is easy to reduce a fraction to its lowest terms when the bottom number is a multiple of 10. Such numbers as 10; 100; 1,000; and so on, can be factored into combinations of only 5s and 2s. Thus, if the top number in such a fraction is not divisible by 2 or 5, the fraction is already in simplest form. The top number must be even or end in 5 if it is to be reduced.

Rule

To change a terminating decimal into a fraction, write a fraction with the decimal as the top number and 1 as the bottom number. Multiply the top and bottom numbers by a multiple of 10 great enough to eliminate the decimal point in the top number.

Try the sample problems. You'll find them easier than you think.

PROBLEMS

Change the following decimals to fractions; then reduce the fractions to simple form.

1. 1.15

2. 2.45

3. 10.01

4. 17.17

5. 1.017

6. 9.8220

ANSWERS

1. $1.15 = \frac{1.15}{1} = \frac{1.15 \times 100}{1 \times 100} = \frac{115}{100} = \frac{23}{20}$

2. $2.45 = \frac{2.45}{1} = \frac{2.45 \times 100}{1 \times 100} = \frac{245}{100} = \frac{49}{20}$

3. $10.01 = \frac{10.01}{1} = \frac{10.01 \times 100}{1 \times 100} = \frac{1001}{100}$

4. $17.17 = \frac{17.17}{1} = \frac{17.17 \times 100}{1 \times 100} = \frac{1717}{100}$

5. $1.071 = \frac{1.017}{1} = \frac{1.017 \times 1000}{1 \times 1000} = \frac{1017}{1000}$

6. $9.8220 = \frac{9.822}{1} = \frac{9.822 \times 1000}{1 \times 1000} = \frac{9,822}{1,000} = \frac{4,911}{500}$

CHANGING INFINITE, REPEATING DECIMALS TO FRACTIONS

How about infinite, repeating decimals? While there is a comple procedure for changing any infinite, repeating decimal into its exa fraction equivalent, you don't need to learn it. In all my experience i solving real-world problems, I can't remember ever having to chang an infinite, repeating decimal to its exact fraction. You frequently nee to turn fractions into decimals because our money system is a decim; system and because our calculators work with decimals. But changin decimals to fractions is rare. When it does occur, it is with terminatin decimals or with infinite, repeating decimals that have been rounde to become terminating decimals.

Rule

To change an infinite, repeating decimal to a fraction, round the decimal and treat it as a terminating decimal. Change it to a fraction as you would any terminating decimal.

HANDLING MIXED FORMS

⁷hen performing calculations for real problems, you may often en-ounter numbers that are really combinations of different kinds of umbers. However, this should not cause any problems, because you ow know how to change one kind of number into another.

For example, you could have a fraction with a decimal rather than whole number as the top number.

$$\frac{9.233}{17}$$

To reduce this fraction to a pure decimal, simply divide 9.233 by 7. Or if you want to write this as a pure fraction, multiply the top d bottom numbers by 1,000.

You might encounter a fraction that has both a decimal and another action. (You might want to try this one on that niece of yours.)

$$\frac{9.233}{^2/_5}$$

In this example, the fraction ⅖ is the bottom number, whereas the p number is still the decimal 9.233. Again, you can change this to a ıre fraction or a decimal. To change the entire fraction to a decimal, ˙st change 2/5 to a decimal by dividing 2 by 5.

$$\frac{9.233}{^2/_5} = \frac{9.233}{0.4}$$

ıe next step is to divide the two decimals.

$$0.4 \, \overline{\big)\, 9.2\ 3\ 3}$$

$$
4\ \overline{\big)\ 9\ 2.3\ 3}
$$

$$
\begin{array}{r}
2\ 3.0\ 8\ 2\ 5 \\
4\ \overline{\big)\ 9\ 2.3\ 3\ 0\ 0} \\
8\ x\ x\ x\ x\ x \\
\hline
1\ 2 \\
1\ 2 \\
\hline
0\ 3\ 3 \\
3\ 2 \\
\hline
1\ 0 \\
8 \\
\hline
2\ 0 \\
2\ 0 \\
\hline
0
\end{array}
$$

You can also change the mixed fraction into a pure fraction. Th takes a little more manipulation. First rewrite the fraction.

$$
\frac{9.233}{^{2}/_{5}} = \frac{9.233}{1} \div \frac{2}{5}
$$

Next change the fraction with the decimal in the top to a pure fractic by multiplying by 1,000.

$$
\frac{9.233}{1} \div \frac{2}{5} = \frac{9,233}{1,000} \div \frac{2}{5}
$$

The next step is to simply invert the $^{2}/_{5}$ and multiply.

$$
\frac{9,233}{1,000} \div \frac{2}{5} = \frac{9,233}{1,000} \times \frac{5}{2} = \frac{46,165}{2,000}
$$

The last step is to reduce the answer to simplest form.

$$\frac{46,165}{2,000} = \frac{9,233}{400}$$

How did your niece do on that one?

You may also see mixed-form numbers in other combinations: for xample, $0.112 + \frac{2}{3}$ or $\frac{4}{1.2} - 5$.

To handle such combinations, simply change one kind of number to another and carry out the indicated operation. The important oint is not to be confused or frightened when you see such forms. Vith what you now know, you can reduce such numbers to pure orms if necessary. Try some problems to see how much your skills ave grown.

PROBLEMS

hange the following mixed-form numbers into pure decimals. When necessary, und at three decimal places to the right of the decimal point. Because several otions exist when performing reductions, your results may turn out to be slightly fferent from those shown. However, your answers should be the same or very ose. To review the process of rounding numbers see Chapter 16.

1. $\frac{0.04}{2}$

2. $\frac{1/2}{11}$

3. $1 + \frac{2}{3}$

4. $1.22 + \frac{7}{2}$

5. $\frac{3/8}{.55} - 1.44$

6. $\frac{1.47}{4/9} + \frac{2/3}{1.12}$

ANSWERS

$$\frac{0.04}{2} = 2 \overline{\smash{\big)}\ 0.0\ 4} \quad = 0.02$$
$$\begin{array}{r} 0.0\ 2 \\ \underline{4} \\ 0 \end{array}$$

$$\frac{1/2}{11} = \frac{0.5}{11} = 11 \overline{\smash{\big)}\ 0.5\ 0\ 0} \quad = 0.045$$
$$\begin{array}{r} 0.0\ 4\ 5 \\ \underline{4\ 4}\ x \\ 6\ 0 \\ \underline{5\ 5} \\ 5\ \text{(remainder)} \end{array}$$

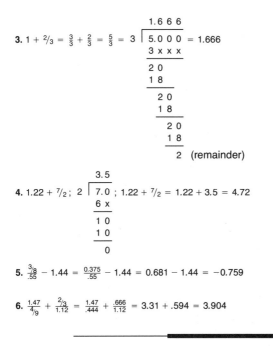

3. $1 + \frac{2}{3} = \frac{3}{3} + \frac{2}{3} = \frac{5}{3} = 3 \overline{)\,5.000} = 1.666$

4. $1.22 + \frac{7}{2}$; $2 \overline{)\,7.0}$; $1.22 + \frac{7}{2} = 1.22 + 3.5 = 4.72$

5. $\frac{\frac{3}{8}}{.55} - 1.44 = \frac{0.375}{.55} - 1.44 = 0.681 - 1.44 = -0.759$

6. $\frac{1.47}{\frac{4}{9}} + \frac{\frac{2}{3}}{1.12} = \frac{1.47}{.444} + \frac{.666}{1.12} = 3.31 + .594 = 3.904$

WHAT HAVE YOU LEARNED?

1. Rule: To change a fraction to a decimal, divide the fraction's to number by its bottom number ($\frac{2}{5} = 2 \div 5 = 0.4$).
2. Rule: To change a terminating decimal to a fraction, write a fractio with the decimal as the top number and 1 as the bottom numbe Multiply the top and bottom numbers by a multiple of 10 grea enough to eliminate the decimal point in the top number.
3. Rule: To change an infinite, repeating decimal to a fraction, roun the decimal and treat it as a terminating decimal.

You have just completed twelve chapters on arithmetic. Where d you stand? If you successfully completed the sample problems at th end of this chapter, you know more about mathematics and can d more arithmetic than most Americans. If you have mastered the mate rial in these first twelve chapters, you are ready to proceed up the stai case to more sophisticated problem solving and mathematics. You hav already done the hardest work. What follows in the next chapters

asier to master and carry out than what you have learned in the
revious chapters.

You can be proud of your new skills. But don't allow yourself to
orget them. Practice the operations you have learned. Use these new
kills in situations that used to freeze you with fear. Use a calculator
when you wish, but try working the problem by hand as well. Then
confirm your answer with your calculator. In this way, you will keep
our skills sharp and your mind toned. Now let's take some more
teps up that staircase.

CHAPTER 13

USING EQUATIONS
TO SOLVE PROBLEMS

In this chapter you'll review equations and go over several simpl laws that can help you manipulate them. You'll also find out how analyze a situation to set up a problem-solving example.

THE BASIC LAWS

An equation is simply a number or set of numbers on the left si of an equal sign and a number or set of numbers of equal value the right side of the equal sign. There are seven basic laws you ca use to manipulate and move numbers around an equal sign. In t following display, letters of the alphabet take the place of numbers.

If A, B, and C are numbers, then

Reflexive law: $A = A$ (e.g., $3 = 3$; there would be few argumen with this one!)

Law of substitution: If $A = B$, then B can be used in place of (e.g., $3 = 2 + 1$; you can use 3 or $2 + 1$ interchangeably.)

Commutative law

For addition: $A + B = B + A$ (e.g., $3 + 2 = 2 + 3$; both sid equal 5.)

For multiplication: $A \times B = B \times A$ (e.g., $3 \times 2 = 2 \times 3$; bo sides equal 6.)

Associative law

For addition: $A + (B + C) = (A + B) + C$; e.g., $3 + (2 + 4)$ $(3 + 2) + 4$; both sides equal 9.

For multiplication: $A \times (B \times C) = (A \times B) \times C$; e.g., $3 \times (2 \times 4)$ $(3 \times 2) \times 4$; both sides equal 24.

Distributive law: $A \times (B+C) = (A \times B) + (A \times C)$; e.g., $2 \times (3+4) = (2 \times 3) + (2 \times 4)$; both equal 14.

Law of equal addition and subtraction: If $A = B$, then

For addition: $A + C = B + C$; e.g., $A = 3$ and $B = 2 + 1$; so $3 + 4 = (2 + 1) + 4$; both equal 7.

For subtraction: $A - C = B - C$; e.g., $3 - 1 = (2 + 1) - 1$; both equal 2.

Law of equal multiplication and division: If $A = B$, then

For multiplication: $A \times C = B \times C$; e.g., $3 \times 1 = (2+1) \times 1$; both sides equal 3.

For division: If C is not equal to zero, $\frac{A}{C} = \frac{B}{C}$; e.g., $\frac{3}{4} = \frac{(2+1)}{4}$; both sides equal ¾.

The above laws date back to the earliest records of mathematics. ꞁe ancient Babylonian civilization, which flourished between 3000 ⸱. and 600 B.C., has provided us with thousands of written records on ay tablets that contain much information about Babylonian math- ꞁatics. The oldest significant mathematical tablets come from the ꞉riod between 2000 B.C. and 1700 B.C., and demonstrate that the ꞁcient Babylonians knew the law of equal addition and subtraction. The ancient Greeks were likewise aware of these laws. *Elements,* ꞉rhaps the most important math book of all time, was written by ꞁclid, a Greek mathematician. Euclid taught at the famous School Alexandria, established in Egypt around 320 B.C. *Elements,* which divided into thirteen books, is the second oldest surviving Greek ath manuscript. It was used as a text for geometry for two thousand ars and influenced mathematics more than any other book. In the ry first book of *Elements,* Euclid states the law of equal addition ꞁd subtraction. In later books he gives the geometrical equivalent ꞏr the distributive law and the commutative and associative laws of ultiplication.

Let's quickly review each of the laws. It's not important that you ꞁemorize the names of each. But it is helpful if you memorize the pression that illustrates each law and understand what each law ꞁeans. First is the reflexive law, which says that for the relationship equality, each number is equal to itself. You express this as $A = A$. It seems obvious that a number is equal to itself. However, other re- ꞁionships exist which do not have the reflexive property. In addition the relationship of equality ($=$), you can express the property of

inequality. For example, to say that 2 is less than 4, you write: 2 <
This is read as "2 is less than 4." The reflexive law does not hold f
inequality. In other words, 2 is not less than itself.

Next you have the law of substitution. This law is also intuitive
obvious and states that if the number A is equal to the number
then you can substitute B for A and A for B. This law is very usef
for simplifying equations; you can substitute simpler expressions f
more complicated ones. Consider the following equation.

$$X = \frac{(5 \times 1) \times (7 \times 1)}{(7 + 5)}$$

You know that $5 \times 1 = 5$, that $7 \times 1 = 7$, and that $7 + 5 = 1$
You can substitute these values in the equation.

$$X = \frac{5 \times 7}{12}$$

The third law is the commutative law for addition and multip
cation. This law states that when you add or multiply two nur
bers, it does not matter in what order the two numbers are. Henc
$A + B = B + A$ and $A \times B = B \times A$. We used this law when memorizi
the addition and multiplication tables.

The fourth law is the associative law for addition and multiplic
tion, which says simply that it does not matter in what order y
add or multiply groups of numbers. The law is illustrated by the tv
expressions $A + (B + C) = (A + B) + C$ and $A \times (B \times C) = (A \times B) \times$
The first expression states that if you have three numbers you can a
them in different orders. If $X = 2 + 5 + 8$, you can first add 2 and
to get 7 and then add 8 to get 15. On the other hand, you could fi
add 5 and 8 to get 13 and then add 2 to get 15.

The associative law for multiplication is just as easy. If $80 = 2 \times$
$\times 8$, you can multiply 2×5 to get 10, which is then multiplied
8 to get 80. Or you can multiply 5×8 to get 40 and then multip
40×2 to get 80. An easy way to remember the name of this la
is to remember that it allows numbers to "associate" with each oth
differently when adding or multiplying.

The fifth law is the distributive law, which is illustrated by $A \times (B$
$C) = (A \times B) + (A \times C)$. One way to remember this law is that
allows you to "distribute" a number inside of the parentheses. Hen

is distributed inside and multiplied with the B and C. An example would be

$$3 \times (4 + 5) = 3 \times 4 + 3 \times 5 = 12 + 15 = 27$$

or

$$3 \times 4 + 3 \times 5 = 3 \times (4 + 5) = 3 \times 9 = 27.$$

The sixth law is the law of equal addition and subtraction, which states that you can add or subtract the same number or numbers on both sides of the equal sign. So, if $A = B$, then $A + C = B + C$, and $A - C = B - C$. Here is an illustration.

$$12 = (7 + 5)$$

$$12 - 6 = (7 + 5) - 6$$

$$6 = 6$$

Although it is true that you have changed the value on either side of the equal sign by subtracting 6, you still end up with *equal* values on either side. In other words, the expression shown is still true.

The seventh law is the law of equal multiplication and division. It states that numbers multiplied or divided on both sides of the equal sign yield equal expressions. Hence, if $A = B$, then

$$A \times C = B \times C \qquad \text{and} \qquad A/C = B/C.$$

There is only one exception to this rule. In the case of division, you cannot use the number 0 because you cannot divide by 0. You should also avoid multiplying by 0, as it results in the trivial case of $0 = 0$. Changing an equation to $0 = 0$ does not help you. Here are examples of the law of equal multiplication and division.

$$12 = 5 + 7$$

therefore

$$\frac{12}{3} = \frac{(5 + 7)}{3}$$

and

$$12 \times 3 = (5 + 7) \times 3$$

By applying these seven laws, you can solve many cumbersom and confusing problems quickly and easily. Use the reflexive law t set up the original equation and then use the other laws to play wit one side of the equation until it yields the solution you seek. Try th sample problems.

PROBLEMS

Identify each statement as true or false according to the laws you have just learne When reducing equations which contain parentheses, carry out the indicated oper tion within the parentheses first and then carry out the other operations.

1. $7 \times (2 + 3) = (7 \times 2) + 3$

2. $14 + 3 = 3 + 14$

3. $62 - 1 = (59 + 3) - 1$

4. $41 \times 2 = (2 \times 40) + 1$

5. $3 \times (9 \times 2) = (3 \times 9) \times 2$

6. $96 = 96$

7. $(7 + 9) + 2 = (2 + 7) + (2 + 9)$

8. $\frac{14}{9} = \frac{(2 \times 7)}{9}$

9. $16 + 6 = 4 \times 4 + 6$

10. $29 \times 2 = (28 + 1) \times 2$

ANSWERS

1. False; you could have used the distributive law to write $7 \times (2 \times 3) = (7 \times 2) + (7 \times 3)$.

2. True; follows the commutative law (addition).

3. True; because of the law of equal subtraction.

4. False; you could have used the commutative law (multiplication) to write $41 \times 2 = 2 \times 41$.

5. True; follows the associative law (multiplication).

6. True; follows the reflexive law.

7. False; mixes up the associative law for multiplication with the associative law for addition—should ha written $(7 + 9) + 2 = 7 + (9 + 2)$.

8. True; follows the law of equal division.

9. True; follows the law of equal addition.

10. True; follows the law of equal multiplication.

MANIPULATION MADNESS

Now that you have reviewed the basic laws used in solving problems, let's go through some fun examples. We'll proceed slowly so that each step is clearly demonstrated. A major part of solving many everyday problems is to take a complicated expression that involves numbers and the four operations of arithmetic and then reduce that expression to one number. You'll encounter this in the use of formulas and in many application problems.

Our first problem is not difficult. You are using a recipe that requires $1\frac{1}{2}$ cups of milk and serves two people. You want to serve 3 people: you, your aunt, and your uncle. You realize that you'll get the correct total if you divide $1\frac{1}{2}$ cups in half to get one serving, and then multiply by 3 people. Dividing by 2 and multiplying by 3 is equivalent to multiplying by $\frac{3}{2}$, which reduces the number of steps you have to take. Begin by using the reflexive law to state the problem. On each side of the equal sign, multiply $1\frac{1}{2}$ cups by $\frac{3}{2}$.

$$(1\,^1/_2) \times (^3/_2) = (1\,^1/_2) \times (^3/_2)$$

To solve this problem, you want to reduce the expression on the right side to one number. Let's begin with the $1\frac{1}{2}$ inside the parentheses. You know that you can multiply the whole number by the bottom number in the fraction, add that to the top number, and put the sum over the original bottom number to make a pure fraction out of a mixed number such as $1\frac{1}{2}$.

$$(1\,^1/_2) \times (^3/_2) = (^3/_2) \times (^3/_2)$$

Now you can multiply the fractions on the right.

$$(1\,^1/_2) \times (^3/_2) = \frac{(3 \times 3)}{(2 \times 2)}$$

Here's your answer.

$$(1\,^1/_2) \times (^3/_2) = {}^9/_4$$

But how in the world do you measure out ⁹⁄₄ cups of milk? Simpl
Remember that ⁹⁄₄ is another way of expressing 9 ÷ 4. Rewrite th
right side of the equation this way.

$$(1\tfrac{1}{2}) \times (\tfrac{3}{2}) = 9 \div 4 \qquad \text{or}$$

$$4 \overline{)\,9}\quad \begin{array}{r}2\\8\\\hline 1\end{array}\text{ (remainder)}$$

Put the remainder 1 into a fraction over the 4, and you have 2¼.

You need 2¼ cups of milk. If this example seemed overly simp
and you could see the answer coming, that's good. It means yo
understand these laws and how they work. On the other hand,
you feel confused, go back to the beginning of the example and, wit
pencil and paper, copy down each equation and follow how each ste
was built. Then try the sample problems.

PROBLEMS

Reduce the right side of each equation to a single number.

1. $(1 - 5) = (1 - 5)$

2. $^{14}/_7 = ^{14}/_7$

3. $^{(4 - 2)}/_2 = ^{(4 - 2)}/_2$

4. $^1/_4 + 7 = ^1/_4 + 7$

5. $\frac{9\times(3+2)}{15} = \frac{9\times(3+2)}{15}$

6. $\frac{(9-4)-5}{11} = \frac{(9-4)-5}{11}$

ANSWERS

1. $(1 - 5) = (-4)$

2. $^{14}/_7 = 2$

3. $^{(4 - 2)}/_2 = 1$

4. $^1/_4 + 7 = ^{29}/_4$

5. $\frac{9\times(3+2)}{15} = 3$

6. $\frac{(9-4)-5}{11} = 0$

INTRODUCING MR. X

efore proceeding, let's introduce two simple conventions to reduce
ır work and make life easier. With the example given in the previous
ction we used the reflexive law to set up the problem as an equation
ith both the right and left sides showing the identical expression. As
ou may know, mathematicians are notoriously lazy and hate to do
xtra work. Instead of writing the same expression on both sides of
ıe equal sign, they just write *X* on the left side. This *X* is the answer
ou are looking for, and *X* is called the unknown. Hence, the equation
ɔm the previous section might look like this.

$$X = (1 + \tfrac{1}{2}) \times (\tfrac{3}{2})$$

"But wait," you say. "If you want me to use an *X* in problem
lving, you're talking about algebra! You didn't say we were going to
ɔ algebra." That's true; I didn't say it. But we are in fact now doing
gebra. The algebra we will be doing is easy to follow and will greatly
mplify solving problems.

The second convention will avoid costly mistakes. In arithmetic, we
present the operation of multiplication with an $\times$, as in $3 \times 4 = 12$.
ɔwever, we can no longer do this, because we are going to introduce
r. *X* into our equations as a number. We don't want to confuse him
ith the symbol for the operation of multiplication. Therefore, we will
e the following convention: When we wish to show two numbers
ultiplied together, such as 3 and 4, we will write a dot between them,
we will enclose both numbers within parentheses. Hence, $3 \cdot 4 =$
, and $(3)(4) = 12$. By enclosing both numbers in parentheses, we
ill avoid confusing 3 times 4 with the number 34. However, when we
e letters or a letter in combination with a number to represent the
ɔduct of multiplication, we don't need a symbol between the letters
between the number and the letter. With these new conventions,
e old arithmetic expression on the left becomes the new algebraic
pression on the right.

$$3 \times 4 = 3 \cdot 4 \qquad \text{(multiplying two numbers)}$$
$$3 \times 4 = (3)(4) \qquad \text{(multiplying two numbers)}$$
$$3 \times A = 3A \qquad \text{(multiplying a number and a letter)}$$

$$A \times A = AA \quad \text{(mulitpling two letters)}$$
$$A \times B = AB \quad \text{(multiplying two letters)}$$

The above shorthand eliminates extra steps and makes our work easier. We will generally show division of A by B as A/B.

The simplest form of an algebraic equation has X all alone on the left of the equal sign, and an expression consisting of several numbers on the right. To find the value of X, the right side of the equation must be simplified.

For example, suppose your living room is a rectangle, and you want to know its area. You measure its length to be 14 feet and its width to be 12 feet. The area of a rectangle is equal to the length times the width. You can now express this as an algebraic equation.

$$X = (\text{length})(\text{width})$$

On the left you have Mr. X, the unknown value you are seeking, or the area of the room. On the right, you show the length times the width. You can now write the equation.

$$X = 14 \text{ ft} \cdot 12 \text{ ft}$$

or

$$X = 168 \text{ square feet}$$

When we multiply feet $\times$ feet, the answer is expressed in square feet. The same is true of other measurements such as inches, miles and yards.

Try the sample problems.

PROBLEMS

Solve each equation for X by reducing the right side of the equation to a single number.

1. $X = 13 - 4$

2. $X = (6 - 1)(4)$

3. $X = {}^{(3 \cdot 2)}/_{17} - {}^{16}/_{17}$

4. $X = {}^{(9+2)}/_{4}$

5. $X = {}^{(3)(5)}/_{15} - 0.5$

6. $X = (4) \times {}^{(7)}/_{2} + 5$

ANSWERS

. *X* = 9   4. *X* = $^{11}/_4$

. *X* = 20   5. *X* = 0.5

. *X* = ($^{-15}/_{17}$)   6. *X* = 19

ISOLATING MR. X

ɔu won't always find equations written with *X* conveniently isolated ll by itself on the left. Sometimes other terms are on the left, and ɔu must do some operations to get *X* by itself. Consider this puzzle. ᴵy father's age plus 1 is 18 times my youngest daughter's age. My ɔungest daughter is 5; how old is my father? The problem is easy ▸ set up. Call my father's age *X*, because this is the unknown that ɔu want to find out. You know that my father's age plus 1 is equal ▸ something else. So, on the left side of the equation, write *X* (my ther's age) plus 1.

$$X + 1 =$$

Now you need to write something on the right side. My father's ϝe plus 1 is equal to 18 times the age of my youngest daughter, who 5. The right side of the equation must be 18 times 5.

$$. X + 1 = (18)(5)$$

You now have the proper equation, but *X* does not stand alone on ▸e left. You can simplify the right side by multiplying 18 × 5 to get).

$$X + 1 = 90$$

How in the world are you going to get *X* alone on the left side? ᴉe solution is to subtract 1 from both sides of the equation. Remem- ▸r that the law of equal addition and subtraction says that subtracting the same number from both sides of an equation does not change th equality. Subtract 1 from both sides.

$$(X + 1) - 1 = 90 - 1$$
$$X + (1 - 1) = 90 - 1$$
$$X + 0 = 90 - 1$$

That gives you the answer.

$$X = 89$$

My father's age is 89. Now that you've worked through the prob lem the long way, you can take shortcuts. Notice that subtractin the number 1 from both sides of the equation $X + 1 = 90$ produce $X = 90 - 1$.

This suggests that you can simply move a number from one sid of an equal sign to the other by changing its operational sign. Yo must be careful that the term you're moving stands alone and is no combined in multiplication or division with other numbers.

Rule
If a term stands alone on one side of an equation, it can be moved to th other side by changing its operational sign.

Consider the following equation.

$$X - 44 = 366$$

Rather than going through all the steps of adding 44 to both side and then regrouping and adding, just move the 44 from the left to th right, and instead of subtracting it, add it.

$$X = 366 + 44$$

Here are some more shortcuts for manipulating equations.

Rule
The two sides of an equation can be exchanged across the equal sig.
$(9 - 2)(14) = X$ is the same as $X = (9 - 2)(14)$

Rule

The positive or negative signs can be simultaneously switched on both sides of an equation. $(-X) = 47$ is the same as $X = (-47)$

Rule

Both sides of an equation can be simultaneously inverted. $1/X = 2$ is the same as $X = \frac{1}{2}$

The next step occurs when X is multiplied or divided by a number. Consider the following equation.

$$3X = (10 + 41)$$

You don't want to know what 3 times X is; you want to know what X is by itself. Apply the law of equal division, dividing both sides of the equation by 3.

$$3X/3 = (10 + 42)/3$$

Notice that when you divided the right hand side of the equation, you divided the entire expression. This is important. You can cancel the 3s on the left to get $X/1$, or just X.

$$X = (10 + 42)/3$$

Now you can use simplification.

The basic question to ask at each step in isolating X is, "What can I do to each side of the equal sign to simplify the left side and help isolate X?" To isolate X when it is the top number of a fraction you do the opposite of what you do when X is multiplied by a number. Consider the following problem.

$$X/3 = 7$$

Multiply each side by 3.

$$(3)(X/3) = (3)(7)$$

On the left, change the 3 into a fraction by making it $^3/_1$.

$$(^3/_1)(^X/_3) = (3)(7)$$

Multiply.

$$^{3X}/_3 = (3)(7)$$

Now you can cancel the 3s on the left side.

$$^X/_1 = (3)(7)$$

Of course, $^X/_1$ is simply X.

$$X = (3)(7)$$
$$X = 21$$

Applying the rules and laws you've learned, try to isolate X in th
following equation.

$$2 = \frac{14}{(1 - 3X)}$$

How do you get X out from the right side to stand alone on th
left? Actually, it's not as hard as it looks. First get X out of the bottom
number of the fraction on the right by inverting both sides (afte
changing the left side from 2 to $^2/_1$).

$$^2/_1 = \frac{14}{(1 - 3X)}$$

becomes

$$^1/_2 = \frac{(1 - 3X)}{14}$$

Now get the 14 out of the bottom of the right side by multiplying both sides by 14. The 14s on the right cancel each other.

$$(\tfrac{1}{2})(14) = \frac{(1 - 3X)}{1} = (1 - 3X)$$

Now multiply the numbers on the left. $(\tfrac{1}{2})(14) = \frac{14}{2} = 7$

$$7 = 1 - 3X$$

Then you can simply exchange both sides.

$$1 - 3X = 7$$

Rewrite the left side.

$$1 + (-3X) = 7$$

Move the 1 to the right.

$$(-3X) = 7 - 1 = 6$$

Now just change the signs of both sides.

$$3X = -6$$

Finally, you can divide both sides by 3 to eliminate the 3 on the left.

$$X = -2$$

You can check your answer by filling the -2 into the original equation in place of the X.

$$2 = \frac{14}{1 - 3(-2)} \quad 2 = \frac{14}{1 - (-6)} \quad 2 = \frac{14}{1 + 6} \quad 2 = \frac{14}{7} \quad 2 = 2$$

You know that your answer of $X = -2$ was correct.

   You have now succeeded in isolating and solving for X. You worked through many steps to do it, but each step was simple and easy.

At first, solving these kinds of problems takes time because you have to be careful and confirm that each change is allowed under the laws of arithmetic and algebra. As your experience grows, so will your speed. When you try the sample problems, think of them as a game, not as a test. You may lose the game on occasion, but that's not a catastrophe. No one is judging you or looking over your shoulder. Work the problems and see how many you can solve correctly. If you miss a few, then you need a little more practice. If you can get them all right, you've mastered a mathematical technique that is beyond most of your contemporaries.

PROBLEMS

Isolate X on the left, and then simplify the right to solve for X.

1. $-X = 3 + 7$

2. $X + 1 = {}^{15}/_7$

3. $2X - 4 = 13$

4. $1/X = {}^4/_{13}$

5. $\frac{1}{(3X+2)} = {}^1/_2$

6. $17 = X - 1$

7. $5 = \frac{1}{(5X-1)}$

8. $9 = \frac{3}{(2+X)}$

9. $\frac{(13-2)}{44} = \frac{1}{(5-2X)}$

10. $2X - 3 = 16 + X$

ANSWERS

1. $X = -10$

2. $X = 8/7$

3. $X = 17/2$

4. $X = {}^{13}/_4$

5. $X = 0$

6. $X = 18$

7. $X = 6/25$

8. $X = (-{}^5/_3)$

9. $X = 1/2$

10. $X = 19$

WHAT HAVE YOU LEARNED?

You have learned the following basic laws and shortcut rules for manipulating numbers around an equal sign.

1. Reflexive law: $A = A$ or $(3 = 3)$
2. Law of substitution: If $A = B$, then B can be used in place of A, and A can be used in place of B.
3. Commutative law
   For addition: $A + B = B + A$ or $3 + 4 = 4 + 3$
   For multiplication: $A \times B = B \times A$ or $3 \times 4 = 4 \times 3$
4. Associative law
   For addition: $A + (B + C) = (A + B) + C$, or $1 + (2 + 3) = (1 + 2) + 3$
   For multiplication: $A \times (B \times C) = (A \times B) \times C$, or $1 \times (2 \times 3) = (1 \times 2) \times 3$
5. Distributive law: $A \times (B + C) = A \times B + A \times C$, or $3 \times (4 + 5) = 3 \times 4 + 3 \times 5$
6. Law of equal addition and subtraction: If $A = B$, then
   For addition: $A + C = B + C$
   For subtraction: $A - C = B - C$
7. Law of equal multiplication and division: If $A = B$, then
   For multiplication: $A \times C = B \times C$
   For division: If C is not 0, then $A/C = B/C$
8. Rule: If a term stands alone on one side of an equation, it can be moved to the other side by changing its operational sign.
9. Rule: The two sides of an equation can be exchanged across the equal sign.
10. Rule: The plus or minus signs can be simultaneously switched on both sides of an equation.
11. Rule: Both sides of an equation can be simultaneously inverted.

To solve a problem, you can use the symbol X to represent the unknown quantity. After setting up the problem, isolate X on the left side of the equal sign and simplify the expression on the right side. You'll see in the next chapter how this useful technique can be applied to percentages.

CHAPTER 14

PERCENTAGES

WHAT ARE PERCENTAGES?

You have already encountered percent signs and percentages in real-world situations. But when you encounter them, do you get a strange feeling in the pit of your stomach? Do you want to run, screaming, in the opposite direction? Percentages sound so simple, but they always seem to get out of hand. Why are percentages necessary, anyway? The answer is simple. Percentages are used to compute your home mortgage, your income tax, sales tax, your interest on savings, discounts at stores, and much more.

Before learning to manipulate percentages, take a moment to consider exactly what they are. A percentage is just another way of writing a number as parts of a hundred. You've seen that a decimal is just another way of writing fractions. In the same way, percentages are just another way of writing fractions or decimals.

Rule
Every fraction or decimal can be written as a percentage, and every percentage can be written as a fraction or a decimal.

Some math books insist that percentages are not numbers but ratios. Although this may be true, it is not necessary information for our purposes. In fact, a percentage *can* be thought of as a ratio, but so can fractions and even decimals. However, we will treat percentages just like ordinary numbers.

Consider the following ways of writing fractions, decimals, and percentages.

$$\frac{1}{2} \; = \; 0.5 \qquad\qquad = 50\%$$
$$\frac{1}{3} \; = \; 0.333\ldots \qquad = 33\tfrac{1}{3}\%, \text{ or } 33.33\%$$
$$\frac{3}{7} \; = \; 0.4285\ldots \qquad = 42.85\%$$
$$11\tfrac{1}{4} \; = \; 2.75 \qquad\qquad = 275\%$$

Notice that with decimals it is sometimes acceptable to use trailing periods to indicate a repeating, infinite decimal. We generally do not use the trailing periods with percentages, but simply round the number. Hence, the decimal 0.333 . . . is written as 33.3% or 33% or 33.333%. The number of digits used is up to you.

CHANGING DECIMALS AND FRACTIONS TO PERCENTAGES

The next issue is how to change a fraction or a decimal to a percentage. It's very simple.

Rule

To change a decimal to a percentage, multiply by 100 and attach the percent sign (%) at the right.

When you multiply by 100, all you have to do is move the digits two places toward the left of the decimal point.

$$16.112 = 1,611.2\%$$
$$.001 = 0.1\%$$
$$3 = 300\%$$
$$0.01961 = 1.961\%$$

In some cases, you write one or more zeros or delete unnecessary zeros. Since you multiply by 100 to find percentages, you can think of a percentage as parts of 100. In other words, 16% is 16 per hundred. If that 16% is a tax, this means you pay 16 cents for every 100 cents, which is the same as 16 cents for every dollar.

Rule

To change a fraction to a percentage, simply convert the fraction to a decimal by dividing; then multiply by 100 and attach the percent sign.

Suppose your son's class contains 20 students, 15 of whom elect to go on a field trip. Those 15 students make up ¾ of the class. What percentage of the class has elected to go on the field trip? Follow the easy steps to convert ¾ to a decimal, 0.75; then write the percentage.

$$\frac{3}{4} = 3 \div 4 = 0.75 = 75\%$$

You can easily see that 75% of your son's class is going on the trip. This should be a snap; try the sample problems.

PROBLEMS

Change each decimal or fraction to a percentage.

1. 0.23	**6.** ½
2. 10.441	**7.** ⅗
3. 17	**8.** 11⁄9
4. 1,001	**9.** 10½
5. 0.002	**10.** 143⁄33

ANSWERS

1. 23%	**6.** ½ = 0.5 = 50%
2. 1,044.1%	**7.** ⅗ = 0.6 = 60%
3. 1,700%	**8.** 11⁄9 = 1.222... = 122.2%
4. 100,100%	**9.** 10½ = 50.5 = 5,050%
5. 0.2%	**10.** 143⁄33 = 4.333... = 433.3%

CHANGING A PERCENTAGE TO A DECIMAL OR A FRACTION

You can also convert any percentage to a decimal or a fraction if you wish.

Rule
To change a percentage to a decimal, delete the percent sign and divide by 100. To change a percentage to a fraction, first change it to a decimal and then change the decimal to a fraction.

Of course, dividing by 100 has the effect of moving the decimal point two positions to the left. Examine the following.

$$43.5\% = (43.5)/100 = 0.435 = 87/200$$

$$101.14\% = (101.14)/100 = 1.0114 = 5,057/5,000$$

$$50\% = (50)/100 = 0.50 = 1/2$$

$$200\% = (200)/100 = 2 = 2/1$$

You can add, subtract, multiply, and divide percentages, but it is important to be careful with these operations. It seems clear that 10% of 50 plus 20% of 50 is 30% of 50. However, 10% of 50% is not 500%, but 5%. Suppose you know that 50% of your savings account is budgeted for amusement, and you want to take out 10% of that 50%. What percentage are you going to take out? (10% = 0.1; 50% = 0.5; 0.1 × 0.5 = 0.05. This decimal equals 5%—simple!) When you multiply and divide percentages, it is wise to convert them first to decimals, and then carry out the operation. Try the sample problems.

PROBLEMS

Change the following percentages into decimals.

1. 1.0%

2. 44.2%

3. 0.11%

4. 101.7%

5. 1,010%

6. 0.0033%

ANSWERS

1. 0.01   **4.** 1.017

2. 0.442   **5.** 10.1

3. 0.0011   **6.** 0.000033

THE BASIC EQUATION

If converting fractions and decimals to percentages is so easy, then why do problems involving percentages seem so difficult? When you encounter a problem that begins, "Compute the percentage of . . . ," do you suddenly get butterflies in your stomach? Do your hands tremble? Do you nearly fall out of your chair in defeat? These reactions are not unwarranted, because problems involving percentages range from the very simple to the extremely complex. When you are asked to compute a percentage, a warning flag *should* go up. Although you don't need a special method for computing a percentage, you do need a way to recognize what the problem is really asking.

Here's an example of the simplest kind of percentage problem, which asks what part of *A* is *B*? This kind of problem usually doesn't provide any surprises. In this kind of problem, you are told that you have a group of items and a subset of that group. What percent of the whole set does the subset represent?

Suppose that there are 50 students at a dance and that 31 of them are girls. What percent of all the students are girls? An equivalent question is, what fraction of the total number of students are girls? Build the fraction by writing the number of girls as the fraction's top number and the number of students as the fraction's bottom number, or 31/50. You know how to change this to a percentage: 31/50 = 0.62 = 62%. So the 31 girls constitute 62% of the 50 students at the dance.

Any problem that asks you to compare a subset of a group with the entire group is solved the same way. Write a fraction. Change the fraction to a decimal, and finally change the decimal to a percentage.

A more difficult type of problem asks you to compare two different items when one is not a subset of the other. In such cases, you frequently encounter percentages in excess of 100. For example, cost of living may have increased to 112% of what it was two years ago. Some people become confused by percentages that exceed 100% because they learned the more restrictive definition of a percentage as a part to a whole. They ask, "How can you have more than 100% of anything?" However, you must remember that percentages are just rewritten fractions and decimals; to have a fraction greater than 100% is not unusual.

The first step in setting up this kind of percentage problem is to determine the number you will use as the base. The base becomes the bottom number in the fraction you are building, the fraction that you will change to a percentage. The item you compare with the base is the fraction's top number. To determine which is the top number and which is the bottom number requires a careful reading of the question.

Look at an example. Expressed as a percent, how does the distance from Mars to the sun compare with the distance from Earth to the sun?

What is being asked here? The problem wants a comparison of the distance between Earth and the sun and the distance between Mars and the sun. That seems obvious. The problem wants this comparison to be shown as a percentage. The distance from Earth to the sun is 93,000,000 miles, and the distance from Mars to the sun is 142,000,000 miles. You should know by now that you're going to have to use these two numbers in a fraction, but which is the top and which is the bottom?

Carefully rereading the problem, you realize that the distance from Mars is being compared to Earth's distance, that is, " ... how does the distance from Mars to the sun compare with the distance from Earth ... " Therefore, Earth's distance is the base. Write Earth's distance as the bottom number and the distance from Mars as the top number.

$$\frac{\text{Mars's distance from sun}}{\text{Earth's distance from sun}} = \frac{142,000,000}{93,000,000}$$

First change the fraction on the right to a decimal.

$$\frac{142,000,000}{93,000,000} = 1.527$$

Now just multiply by 100 and attach the percent sign.

$$1.527 = 152.7\%$$

Therefore, if the distance from Earth to the sun is considered 100%, then the distance from Mars to the sun is 152.7%. This example demonstrates the utility of percentages. Looking at the percentages, you can see that Mars is more than 50% farther from the sun than Earth is. It's easier to visualize the distance from Mars to the sun with percentages than with the raw mileage, since the numbers are so large.

Let's use this same example to illustrate a more complex type of percentage problem. Expressed as a percentage, how much farther is Mars from the sun than Earth is from the sun? It looks like the same question, but it's not! The difference is in the wording of the problem and is given away by the words *how much farther.* This is a question not of simply comparing two distances but of comparing a *difference* of distances. Whenever you're working a percentage problem, you need two numbers to put into a fraction. The two numbers you have so far for this problem are the distances from Earth to the sun and Mars to the sun, 93 million miles and 142 million miles. But these two numbers alone will not provide the answer.

In a difference problem, you have to subtract the base from the other number. In this case, 93 million miles is the base. Subtract 93 million from 142 million. Mars is 49 million miles farther from the sun than Earth is. What percentage does this represent? In other words, what percent of 93 million is 49 million? Now you have the two numbers necessary to build the fraction: (49 million)/(93 million) = 0.527. Convert this decimal to a percentage to get 52.7%. Mars is 52.7% *farther* from the sun than Earth is.

Whenever you encounter a percentage problem with the words *farther, greater, larger,* you must read the problem carefully and verify that it is a problem that requires subtraction. Consider how you would set up the following questions if the numbers were provided.

- How much farther is Earth from the sun than Mercury is from the sun?
- Expressed as a percent, how much taller is the Sears Tower in Chicago than the World Trade Center in New York?
- Compared to the population of the United States, how much greater is the population of China?

All three of the above examples require subtracting one number from another before you can build your fraction. The fraction is then used to compute the percentage.

Sometimes you are given a percentage and part of the fraction that made the percentage. Let's look at the three parts of the equation for changing a fraction to a percentage.

$$\frac{\text{Top number}}{\text{Bottom number}} \times 100 = \text{Percentage}$$

Up to now, you have been given the top and bottom numbers and asked to figure the percentage. In many problems, however, you are given either the top number or the bottom number and the percentage, and then asked to compute the part that is missing. For example, you might be given the following problem: At 0.15 miles per hour, a three-toed sloth's speed is 500% of the speed of a garden snail. How fast does a garden snail go? You have two of the three parts of the percentage equation.

$$\frac{\text{Speed of sloth}}{\text{Speed of snail}} \times 100 = 500\%$$

Substitute 0.15 for the speed of the sloth.

$$\frac{0.15}{\text{Speed of snail}} \times 100 = 500\%$$

The first step is to change the 500% to a decimal, or 5.00. To do this, divide both sides of the equation by 100. This cancels the 100 on the left and changes 500 to 5.00. You can now drop the percent sign on the right.

$$\frac{0.15}{\text{Speed of snail}} = 5.00 = 5$$

Remember that *speed of snail* is now X, or the unknown quantity. From Chapter 13, you know how to manipulate an equation. You need to get the snail's speed isolated on the left side of the equation. You can do this in two steps. First invert both sides of the equation.

$$\frac{\text{Speed of snail}}{0.15} = \frac{1}{5}$$

Now multiply both sides by 0.15, which cancels the 0.15 on the left and produces the answer.

$$\text{Speed of snail} = (^1/_5)(0.15) = 0.03$$

The snail's speed is 0.03 miles per hour. You can easily finish this equation before the snail could even cross your garden!

Use the same procedure for any problem in which you are given a percentage and only one of the two numbers used for making that percentage. Set it up as an equation and then solve for the missing number.

Now let's work with a slightly different problem. At 0.15 miles per hour, a three-toed sloth is 400% *faster* than a garden snail. How fast is a snail? See the difference? We're comparing not the *total* speeds but rather the *difference* between the two. This problem involves subtracting; so you must set it up differently.

$$\frac{(\text{speed of sloth}) - (\text{speed of snail})}{(\text{speed of snail})} \times 100 = 400\%$$

In the top number, subtract the snail's speed from the sloth's speed because 400% measures the *difference* in their speeds. You know the speed of the sloth.

$$\frac{(0.15) - (\text{speed of snail})}{(\text{speed of snail})} \times 100 = 400\%$$

Proceed to isolate the speed of the snail on the left side of the equation. Change the 400% to 4.00 by dividing both sides of the equation by 100.

$$\frac{(0.15) - (\text{speed of snail})}{(\text{speed of snail})} = 4.00 = 4$$

Now multiply both sides of the equation by the snail's speed. On the left, this cancels with the snail's speed in the fraction's bottom number.

$$(0.15) - (\text{speed of snail}) = 4 \times (\text{speed of snail})$$

Now move the speed of the snail on the left to the right by adding it to both sides. On the left, it subtracts out.

$$(0.15) = 4 \times \text{(speed of snail)} + \text{(speed of snail)}$$

Add the two terms on the right.

$$(0.15) = 5 \times \text{(speed of snail)}$$

Next switch both sides of the above equation.

$$5 \times \text{(speed of snail)} = (0.15)$$

Now divide both sides by 5.

$$\text{speed of snail} = {}^{(0.15)}/_5 = 0.03$$

Again, 0.03 miles per hour is the snail's speed.

Some people will tell you that all percentage problems are easy, but this is simply not true—percentages can be tough. The math is simple, but figuring out what the problem is about can be difficult. Later in the book, you will explore the application of percentages to interest rates, loans, and mortgages. Try solving some problems.

PROBLEMS

Express the solution to each problem as a percent.

1. Humans have 32 teeth, 12 of which are molars. What percentage of our teeth are molars?
2. Your checking account balance averages $62.50. The bank takes out $4.50 each month for a service fee. What percentage of your average balance is taken out by the bank each month?
3. Compare the gestation period of a goat (151 days) with that of a guinea pig (68 days).
4. How much taller is the Angel Falls waterfall in Venezuela (3,281 feet) than the upper waterfall of Yosemite Falls in California (1,430 feet)? Express your answer as a percentage.
5. The original price of Jack the Joker's 20-inch color television set was $349. Jack is putting the sets on sale at 15% off. What is the cost of a set now?

ANSWERS

1. Percentage = $\dfrac{12 \text{ molars}}{32 \text{ total teeth}}$ = $\dfrac{12}{32}$ = 0.375 = 37.5%

2. Percentage = $\dfrac{\$4.50}{\$62.50}$ = 0.072 = 7.2%

3. Percentage = $\dfrac{151 \text{ days (goat)}}{68 \text{ days (guinea pig)}}$ = 2.22 = 222%

4. Percentage = $\dfrac{\text{(Yosemite Falls)} - \text{(Angel Falls)}}{\text{Angel Falls}}$

Percentage = $\dfrac{3,281 - 1,430}{1,430}$ = $\dfrac{1,851}{1,430}$ = 1.294 = 129.4%

5. Percentage = $\dfrac{\text{(old price)} - \text{(new price)}}{\text{(old price)}}$ = 15%

$\dfrac{349 - \text{(new price)}}{349}$ = 15% = .15

new price = 349 − (.15) × (349) = \$296.65

WHAT HAVE YOU LEARNED?

1. Rule: Every fraction or decimal can be written as a percentage, and every percentage can be written as a fraction or a decimal.
2. Rule: To change a decimal to a percentage, multiply by 100 and attach the percent sign (%) on the right. To change a fraction to a percentage, first convert the fraction to a decimal by dividing. Now multiply by 100 and attach the percent sign.
3. Rule: To change a percentage to a decimal, delete the percent sign, and divide by 100. To change a percentage to a fraction, first change it to a decimal, then change the decimal to a fraction.

The basic percentage problem consists of three numbers: the top number of a fraction, the bottom number of a fraction, and the percentage itself.

$$\frac{\text{top number}}{\text{bottom number}} \times 100 = \text{percentage}$$

If two of the three numbers are known, you can manipulate the equation to solve for the third number.

The next chapter will take you up yet another step of the math staircase—working with powers and roots.

CHAPTER 15

PLAYING WITH

POWERS AND ROOTS

n this chapter you will investigate powers of numbers and roots of numbers. You need to understand powers and roots because they are sometimes found in formulas, especially formulas for computing areas and volumes. You may find the calculator invaluable when you are working with powers and roots.

WHAT ARE POWERS?

Powers of numbers are not difficult to understand. If you wish to multiply the number A by itself, you write it as $A \times A$. There is another way to show this. Instead of repeating the A, you place a small raised 2, or a superscript, to the right of the original A; so, $A \times A = A^2$. To multiply $A \times A \times A$, you write the superscript 3, or $A \times A \times A = A^3$. Hence, the superscript number tells how many times a number is multiplied by itself. This superscript is called an *exponent*. The large number under the exponent is called the *base*. The number resulting from the repeated multiplication is called the *power*. Let's go over these definitions again.

A^n is the *power*.

A is the *base*.

n is the *exponent*.

You will encounter taking A to its nth power, or raising A to its nth power, (A^n). Another common description is to say that raising a number to its second power (e.g., A^2) is *squaring* that number. Hence 5 squared is $5^2 = 5 \times 5 = 25$. When you raise a number to its third power, you are *cubing* that number. So, $2^3 = 2 \times 2 \times 2 = 8$.

The usefulness of this shorthand should be apparent at once. Consider the following product.

$$7 \times 7 \times 7 \times 7 \times 7 \times 7 \times 7 \times 7 \times 7 \times 7 \times 7 \times 7 \times 7 \times 7$$

This number is hard to read because you have to look closely and carefully count the individual 7s. It's easy to make a mistake. However with exponents, you can write the number as 7^{14}. At once you know this number means the number 7 multiplied by itself 14 times, or 7 raised to the 14th power.

If a number stands by itself, you say it is raised to its first power, or $A = A^1$. You can also show an exponent as 0. Any number raised to its zero power is always defined as 1. Hence, $A^0 = 1$, where A is any number.

Definition
$A = A^1$, and $A^0 = 1$.

In applications where you encounter numbers with exponents, the exponents are almost always 2 or 3. This occurs because you often compute areas by squaring numbers and volumes by cubing numbers. Try the sample problems.

PROBLEMS

Compute the power for each base raised to the indicated exponent.

1. 2^2	**6.** 5^3
2. 3^2	**7.** 5^1
3. 2^3	**8.** 4^0
4. 4^2	**9.** 7^3
5. 8^2	**10.** 2^5

ANSWERS

1. 4	**6.** 125
2. 9	**7.** 5
3. 8	**8.** 1
4. 16	**9.** 343
5. 64	**10.** 32

OTHER BASES

So far you have dealt only with bases that are positive whole numbers. But you can also have bases that are negative numbers, fractions, or decimals. It's just a matter of multiplying the base number the appropriate number of times as indicated by the exponent.

Let's try a decimal for a base. What is $(2.3)^2$? It's simply the decimal 2.3 multiplied by itself, or $(2.3)^2 = (2.3) \times (2.3) = 5.29$. If you can multiply decimals, you can do powers with decimal bases. Bases with fractions are just as easy.

$$(2/3)^2 = (2/3) \times (2/3) = (2 \times 2)/(3 \times 3) = 4/9$$

and

$$(2/3)^3 = (2/3) \times (2/3) \times (2/3) = (2 \times 2 \times 2)/(3 \times 3 \times 3) = 8/27$$

In real-world applications, you almost always have to compute powers with bases that are either decimals or fractions. For example, if you want to figure out how much material you need to make a round tablecloth, the formula for calculating the area of a circle is πr^2, where r is the circle's radius (the radius = 1/2 the diameter). The symbol π (the Greek letter *pi*) stands for the ratio of the diameter of a circle to its circumference and is approximated by the decimal number 3.14. If your table has a 3-foot radius (because it has a 6-foot diameter), you can substitute these values to find the circle's area.

$$\text{Area of circle} = \pi r^2 = (3.14) \times (3)^2$$

From the definition of exponents, you know that $(3)^2$ is just 3×3, or 9.

$$\text{Area of circle} = (3.14) \times (9) = 28.26 \text{ square feet}$$

You'll need at least that much material to cover the table.

The volume of a cylinder (or the measurement of the space that the cylinder encloses) is $\pi r^2 H$, where r is the radius of the cylinder's base and H is the cylinder's height. The volume of a sphere is $(4/3)\pi r^3$, where r is the radius.

You can also have bases that are negative numbers. Consider the following power.

$$(-2.3)^2 = (-2.3) \times (-2.3) = 5.29$$

The answer (5.29) turned out to be a positive number because the result of multiplying two negative numbers is a positive number. If the exponent had been 3, you would have found the following:

$$(-2.3)^3 = (-2.3) \times (-2.3) \times (-2.3) = (5.29) \times (-2.3) = (-12.167)$$

In the above case you ended up with a negative answer because, in the last step, you multiplied a positive number (5.29) by a negative number (-2.3) to get a negative result. The trick to raising negative bases is to keep track of the sign of the answer. A simple rule states that, if the exponent is even, the answer will be positive; if the exponent is odd, the answer will be negative. Try some problems.

PROBLEMS

Compute the power for each base raised to the indicated exponent.

1. $(1.1)^2$ 6. $(5/3)^3$

2. $(1/2)^2$ 7. $(10.1)^3$

3. $(3/5)^2$ 8. $(17.13)^1$

4. $(4.7)^2$ 9. $(31/67)^0$

5. $(-3.2)^3$ 10. $(-2.83)^2$

ANSWERS

1. 1.21

2. 1/4

3. 9/25

4. 22.09

5. −32.768

6. 125/27

7. 1,030.301

8. 17.13

9. 1

10. 8.0089

COMPUTING ROOTS

Computing roots is just the reverse of computing powers. With powers, you are given the base and exponent and asked to find the power. With roots, you are given the power and the exponent and asked to find the base.

Begin with the square root. The square root of 4 is 2, which means that 2 × 2 equals 4. In other words, the square root of 4 is a number that when multiplied by itself is equal to 4. How about the square root of 9? It's 3, or 3 × 3 = 9. Does 2 have a square root? Yes. You can't write the square root of 2 out exactly with a decimal or a fraction, but it is approximately the decimal 1.414. In fact, all positive numbers have square roots, or numbers that when multiplied by themselves give you the original number. Numbers can have more than one root. For example, (−2) multiplied by (−2) is +4. This shows that both +2 and −2 are square roots of 4.

There are also cube roots. The cube root of a positive number A is another number that when multiplied by itself 3 times gives you the number A. The cube root of 27 is 3. Why? Because 3 × 3 × 3 = 27. In the same way, the cube root of 8 is 2, or 2 × 2 × 2 = 8. Again, all positive numbers have cube roots. I think you can see where I am leading. Just as there are square roots and cube roots, there are also fourth roots, fifth roots, and so on. For example, the 24th root of 5 is a number that when you multiply it by itself 24 times gives 5 for an answer.

You can write these roots by using the *radix*, $\sqrt{\ }$, to indicate a root. For a square root, you can use the radix by itself or you can include a small 2 above the vee in the radix. So the square root of 3 can be written as either $\sqrt{3}$ or $\sqrt[2]{3}$. When you get to the larger roots, always include a small number above the radix to tell the root's degree. Thus, the 5th root of 9 is written as $\sqrt[5]{9}$. The result will be a number that, when multiplied by itself 5 times, produces 9.

While many roots turn out to be positive whole numbers, this is not always the case. In fact, the roots of most positive numbers are only approximated by decimals or fractions. Consider the following:

$$\sqrt[3]{3} = 1.442$$

$$\sqrt[4]{5.2} = 1.51$$

The answers in the two preceding examples are only approximate answers. You can multiply the numbers on the right to verify that they do approximate the roots. The standard method used to compute roots is too complicated for us to get into here. However, most common applications that require solving for roots involve computing square or cube roots rather than those of higher numbers. So you can use two shortcuts. First, most calculators compute square roots. You just enter the number, press the appropriate button, and the screen shows the number's square root.

The second method is to approximate the answer by guessing. For example, suppose you wanted to build a metal tank in the shape of a cube to hold 800 cubic feet of drinking water. The formula to find the volume of a cube is Volume of cube = (length of a side)^3. Our holding tank is to be 800 cubic feet.

$$800 = (\text{length of side})^3$$

The solution is to find a number representing the length of one side of the tank that when multiplied by itself three times is equal to 800. In other words, you want to find the cube root of 800. First use guesswork. Guess that it is 10 feet.

$$10^3 = 10 \times 10 \times 10 = 1,000$$

The guess of 10 feet was too great. Try 5 feet.

$$5^3 = 5 \times 5 \times 5 = 125$$

This falls far short. The solution must be closer to 10 feet. Try 8 feet.

$$8^3 = 8 \times 8 \times 8 = 512$$

Still off. Try 9 feet.

$$9^3 = 9 \times 9 \times 9 = 729$$

Now you're close—only 71 cubic feet short. Try to get a little closer by testing 9.5 feet.

$$(9.5)^3 = (9.5) \times (9.5) \times (9.5) = 857$$

This gives us a larger number than we want. Try 9.3 feet.

$$(9.3)^3 = (9.3) \times (9.3) \times (9.3) = 804$$

Now you're very close. Do you need more accuracy? If so, make more refined guesses. This method is inexact and can be a little time consuming, but it works, and it is easy to apply. Try some problems.

PROBLEMS

Find the root of each number by using a combination of a calculator and the guessing method. Your answers need not always be exact.

1. $\sqrt[2]{4}$

2. $\sqrt[2]{81}$

3. $\sqrt[2]{54}$

4. $\sqrt[3]{27}$

5. $\sqrt[3]{3.4}$

6. $\sqrt[3]{14.4}$

ANSWERS

1. 2

2. 9

3. 7.35

4. 3

5. 1.5

6. 2.433

WHAT HAVE YOU LEARNED?

You have learned about powers, exponents, bases, and roots. A number, A, multiplied by itself n times is expressed as A^n.

A^n is the *power*.

A is the *base*.

n is the *exponent*.

You also have the following definition: $A^1 = A$, and $A^0 = 1$.

Roots are just the reverse of powers. The nth root of A is a number that when multiplied by itself n times is A. Roots of positive numbers can be estimated by the guessing method. Now let's find out how to use all this.

CHAPTER 16

APPLICATIONS

nowing as much about math as you do, you can begin to apply
our knowledge. The first three sections of this chapter will deal with
nding the unit of measurement of an answer as a means of checking
our work, rounding numbers, and using formulas. The remaining
ctions will provide specific examples.

DETERMINING THE UNIT OF MEASUREMENT

1 real-world applications of mathematics, the answer is generally in
et, inches, pounds, or some other unit of measurement. In many
mple problems, you ignore the unit of measurement, do the calcu-
tions, and then attach the correct unit of measurement when you're
one. For example, you might want to find the number of square feet
your living room. You measure the length and width; they are 14
et and 12 feet. Then set up the problem.

$$\text{Area} = 14 \times 12 = 168$$

fter you're done, attach the unit of measurement.

$$\text{Area} = 168 \text{ square feet}$$

This is a perfectly acceptable procedure, especially for simple prob-
ms. However, in complex problems, when you are dealing with sev-
al different values and units of measurement, you might want to
ork the problem with the units of measurement in order to check

the answer. You can do this by including the units of measureme: with the numbers and carrying out the various operations. The resul ing answer should be in the correct units.

Suppose you take a trip of 144 miles and use 9 gallons of gas. Wh was your gas mileage? You want to know how many miles per gallc you got. Whenever you encounter *per,* the unit of measurement th follows is divided into the previous number. Set up the problem.

$$\text{Mileage} = \frac{144 \text{ miles}}{9 \text{ gallons}} = 16 \text{ miles/gallon}$$

The correct is 16 miles/gallon, or 16 miles per gallon.

Sometimes you may encounter a problem with one unit of mea surement, but you want to express the answer in a different unit (measurement. For example, you would not want the volume of a ter nis ball in cubic miles or the volume of Lake Superior in cubic inche Therefore, you must at times change from one unit to another. This accomplished by substitution.

If you walk a distance of ¼ mile, how many feet have you walked There are 5,280 feet in 1 mile. Therefore, you are going to substitu: 5,280 feet for the mile in ¼ mile.

$$\text{¼ mile} = (\text{¼}) \times (5,280 \text{ feet}) = 1,320 \text{ feet}$$

Try another example. Suppose you know your car is traveling : 55 miles per hour. How fast is it going in feet per second? Conve. miles to feet: 1 mile = 5,280 feet. Convert hours to seconds: 1 hou = 3,600 seconds. Substitute the feet and the seconds for miles an hours.

$$55 \text{ miles/hour} = \frac{55 \text{ miles}}{1 \text{ hour}} = \frac{55 \times (5,280 \text{ feet})}{1 \times (3,600 \text{ seconds})}$$
$$= 80.67 \text{ feet/second}$$

The relationships between various units of measurement chang when you square or cube them. For example, 1 foot is equal to 1 inches. Is 1 square foot equal to 12 square inches? No. Convert b substitution to find how many square inches are in 1 square foot.

1 sq. ft. = (1 ft.) × (1 ft.) = (12 in.) × (12 in.) = 144 sq. in.

So, 1 square foot is really 144 square inches. Try the sample prob-
ns.

PROBLEMS

•mpute the correct answer to each problem, including the correct unit of measure-
?nt. You may not recognize the formulas employed, but that shouldn't matter. Your
erest is in manipulating the numbers and units to arrive at the correct answer.
lve each problem for X.

• $X = \dfrac{55 \text{ miles}}{\text{hour}} \times 3 \text{ hours}$

• $22 \text{ miles/gallon} = \dfrac{X}{17 \text{ gallons}}$

• $X = \dfrac{32 \text{ feet}}{\text{second}^2} \times (10 \text{ seconds})$

. 1 square yard = X square feet   (1 yard = 3 feet)

ANSWERS

$X = 165$ miles (The hours canceled; note that you can cancel equivalent units of measurement just as
you can cancel numbers.)

$X = 374$ miles (The gallons canceled)

$X = 320$ feet/second

9 square feet

ROUNDING NUMBERS

is often convenient to shorten numbers before carrying out calcu-
tions. For example, there is usually no need to compute an answer
volving dollars to more than a penny. You have no need for an an-
ver such as $14.27996. Therefore, you need a method to shorten
umbers without introducing too much error. Up to this point, you

have simply eliminated digits to the right of the decimal point. T]
method known as *rounding* can be used to eliminate excess digits
many situations. You can either round a value down, or round it u

Consider the number 4.12. Suppose you want to use not the enti
number but only the digit to the left of the decimal point, 4. To roun
cut off the 1 and the 2 to get 4. You have lost some accuracy, but n
much. In fact, you have lost only 3% of the original number.

Now consider the number 4.92. You still want only the digit
the left of the decimal point. If you round by cutting off the 9 ar
the 2, you end up with 4 again. But this time you have lost 0.92
the original number, which represents 18.7% or almost one fifth! T]
solution is to round the number up by dropping the 9 and the 2 b
adding 1 to the 4 to get 5. How far off are you if you use 5 instead
4.92? Surprisingly, you are off only 1.6%. An easy way to rememb
this is to think of the number line; 4.92 is closer to 5 on the numb
line than it is to 4.

When you round large numbers with digits to the left of the decim
point, you can't drop digits. Instead, replace the digits with zeros.
you want to round the number 1,824 to retain the two left-most dig
(the 1 and the 8), change the 2 and the 4 to zeros. So, 1,824 becom
1,800. At times, you add 1 to the right-most digit you are keeping.
you round 1,875, you get 1,900.

Following is a rule to tell you when to round up and when to rour
down.

Rule
*If the left-most digit you are eliminating is 5 or greater, add 1 to t
right-most digit being saved. If the left-most digit being eliminated is 4
less, drop it or replace it with 0.*

If you want to round 3.475 to two digits to the right of the de
mal point, you get 3.48 by dropping the 5 and adding 1 to the 7. C
the other hand, if you round 3.474 to two digits to the right of t]
decimal point, you get 3.47 by just dropping the 4. Let's say you a
grocery shopping and have four items in your cart. You don't want
spend more than $15. If the butter costs $3.69, the milk $1.25, t]
hamburgers $6.78, and the buns $1.20, round the numbers to ad
in your head. Have you exceeded your budget? The numbers wou
round to 4 + 1 + 7 + 1 = $13. Your estimation is rough, but yc

ould still be safely within your $15 limit. (The actual sum is $12.92.)
y some rounding samples.

PROBLEMS

und each number, keeping the two left-most digits.

. 3.1099	**4.** 3.19355
. 101.405	**5.** 16.0
. 0.9954	**6.** 18,163

ANSWERS

3.1	**4.** 3.2
100	**5.** 16
1.0	**6.** 18,000

USING FORMULAS

formula is simply a mathematical rule for computing a desired
iantity. You have already worked with certain formulas: The area
a rectangular room is the length times the width; gas mileage is
e number of miles driven divided by the number of gallons of gas
ied. Formulas can be very useful, as they provide a quick and easy
ay to solve a problem without doing a lot of memorizing. All you
ally need to know is where to find the required formula and how
use it.

The easiest formula to use is one that yields the answer directly. For
ample, when you want to know mileage, the formula is set up with
e unknown quantity on the left of the equal sign and the known
lues on the right.

Mileage = (miles driven)/(gallons of gas used)

All you need to do is plug in the known values on the right ar carry out the operation of division. However, this clear-cut situatic does not always exist. Suppose your mileage was 16 miles per gallc and you were going on a trip of 176 miles. How many gallons gas would you need? Now when you plug the known values in the equation, the unknown value (gallons of gas) is not isolated c the left; it is mixed up with a known value on the right. In suc cases you need to manipulate the equation to isolate the unknow on the left and then reduce the right side to a single value. In tl following sections and in the next two chapters, you will be introduce to formulas for solving everyday problems. When the unknown valt is not isolated on the left, don't panic. Simply manipulate the equatic until it is.

ELECTRICAL USAGE

Electrical usage is measured in watts per hour or kilowatts per hou where 1 kilowatt = 1,000 watts. Electrical power is similar to wat flowing in a pipe. Electrical voltage is like the water pressure, where amperage is like the volume of water flowing. You multiply the voltaǥ (pressure) times the amperes (volume) to get watts (energy). Here the basic formula for electrical usage.

$$\text{Watts} = (\text{volts}) \times (\text{amperes})$$

Most of the appliances in your home are rated in watts so it relatively easy to compute how many watts or kilowatts they consun each hour. A standard 75-watt light bulb consumes 75 watts per hou In 10 hours of use, that's 750 watts, or 0.75 kilowatts. At 10 cents p kilowatt hour, the light bulb will cost (10 cents) × (0.75 kilowatts), c 7.5 cents for 10 hours. This illustrates the second formula for electric usage.

$$\text{Cost} = (\text{rate in cents/kilowatt hours}) \times (\text{usage in kilowatt hours})$$

or simply

$$\text{Cost} = (\text{rate}) \times (\text{usage})$$

The typical modern color television set uses remarkably little ectricity—as little as 100 watts (per hour). To run such a TV for) hours consumes 1,000 watts, or 1 kilowatt. At 10¢ per kilowatt our, that will cost you only 10¢—quite a bit cheaper than paying to ·e 10 hours of movies at a theater!

The things in your home that use lots of electrical energy are gener- ly electric heaters, ranges, water heaters, and dryers. A water heater lay consume 5,000 watts per hour, or 5 kilowatts per hour. One our of usage at 10 cents per kilowatt costs 50 cents. A space heater inning at 1,500 watts consumes 6,000 watts in four hours. At 10 :nts per kilowatt, that's 60 cents of electricity.

Sometimes an appliance is rated not in watts or kilowatts but in)lts and amperes. This is common for motors, which generally have uge electrical appetites. A typical large shop motor may be rated at) amperes and 120 volts. To find the wattage, multiply the amperes nd volts.

$$\text{Watts} = (\text{volts}) \times (\text{amperes}) = (120) \times (10) = 1,200 \text{ watts}$$

One hour of use for this motor at 10 cents per kilowatt hour would)st 12 cents.

The electrical company measures the number of kilowatt hours you se each month and computes your bill based on their rates. The)st of residential electricity varies greatly across the country, ranging om as low as 2.9 cents per kilowatt hour (Seattle, Washington) to ; high as 13.9 cents per kilowatt hour (Long Island, New York). he rates can also vary by how much electricity you use, increasing r decreasing with additional energy consumption. Most residential ites are between 6 and 10 cents per kilowatt hour. A homeowner sing 750 kilowatts in Memphis, Tennessee, might pay $43 a month, ·hereas a homeowner in New York City, using 3,000 kilowatts, might e billed $370.

Check your electrical bill the next time it comes. Find out what your ite is per kilowatt hour. Then review your electrical usage to find ·eas where you can conserve and reduce your electrical bill. Simply irning off unused lights and lowering the temperature on your water eater can lower your bill. Try the sample problems. Remember to read ach problem carefully to determine what the real question is.

PROBLEMS

1. Three 60-watt light bulbs are left on 24 hours a day in your garage. If you electrical rate is 8.5 cents per kilowatt hour, what do the 3 lights cost you each year?

2. During 3 months of the winter, you use a space heater rated at 1,200 watts for 7 hours a day. At 10 cents per kilowatt hour, what does this space heater cost each year?

3. Your electrical bill for an average month is $72. Your electrical rate is 7.8 cents per kilowatt hour. How many kilowatt hours do you use in an average month?

4. You have an electric motor that pumps water from your well. It runs 2 hours a day and is rated at 120 volts and 8 amperes. How many kilowatts does the motor use each month?

ANSWERS

1. Three 60-watt bulbs consume 180 watts per hour. Multiply this by 24 hours to get 4,320 watts, or 4.3 kilowatts, per day. Multiply this by 365 days to get 1,576.8 kilowatts per year. The rate is 8.5 cents per kilowatt.

$$\text{Cost} = (8.5 \text{ cents/kilowatt-hour}) \times (1,576.8 \text{ kilowatt-hours}) = \$134.03$$

2. 1,200 watts running for 7 hours per day is 8,400 watts, or 8.4 kilowatts, per day. For 90 days (3 months) this is 756 kilowatts. The rate is 10 cents per kilowatt hour.

$$\text{Cost} = (10 \text{ cent/kilowatt-hour}) \times (756 \text{ kilowatt-hours}) = \$75.60$$

3. Use the standard cost formula.

$$\text{Cost} = (\text{rate}) \times (\text{usage})$$

or

$$\$72 = (7.8 \text{ cents/kilowatt-hour}) \times (\text{usage})$$

Solve.

$$\text{Usage} = (\$72)/(\$0.078) = 923 \text{ kilowatts}$$

4. To begin, use the formula: Watts = (voltage) × (amperage)

$$\text{Watts} = (120) \times (8) = 960 \text{ watts per hour}$$

Running for 2 hours, the motor uses 2 × 960 or 1,920 watts, or 1.92 kilowatts. Over 30 days (1 month) this equals 57.6 kilowatts.

GAS CONSUMPTION

‣u remember the formula for computing gas mileage.

$$\text{Mileage} = \frac{\text{miles driven}}{\text{gallons used}}$$

Of course, you must know two of the items above to compute the ird one. Practice using this formula by solving for the unknown ‣antity in the following problems.

PROBLEMS

‣ mileage $= \dfrac{390 \text{ miles}}{13 \text{ gallons}}$

‣ 26 mpg $= \dfrac{572 \text{ miles}}{\text{gas}}$

‣ 12 mpg $= \dfrac{\text{distance}}{7 \text{ gallons}}$

‣ 37 mpg $= \dfrac{22 \text{ miles}}{\text{gas}}$

ANSWERS

mileage = 30 mpg

gas = 22 gallons

distance = 84 miles

gas = 0.59 gallons

UNIT PRICING

ιit pricing is the practice of computing costs of goods per unit of ‣easurement. This allows you to compare different brands of goods ‣d different sizes of the same brand to find the lowest price. Many grocery stores now show the unit price for each item so that you c compare without doing any arithmetic.

Computing a unit price is easy. Simply divide the price by the vc ume, weight, or other measure of quantity to get a cost per unit measurement. Here's the formula.

$$\text{Unit price} = (\text{price})/(\text{number of units})$$

Suppose you have a 5-ounce bottle of suntan lotion for $3.45; t unit price per ounce is easy to find.

$$\text{Unit price} = (3.45 \text{ dollars})/(5 \text{ ounces}) = \$0.69 \text{ per ounce}$$

The unit price is 69 cents per ounce. Now compare this brand suntan lotion with another brand that sells for $6.44 and contains ounces. Find the unit price for the second brand.

$$\text{Unit price} = (6.44 \text{ dollars})/(7 \text{ ounces}) = \$0.92 \text{ per ounce}$$

You can see at once that the second brand is 23 cents per our more expensive.

If all unit pricing were as simple as the above example, you wot have an easy time finding the best bargains. But, as you can gue things get complicated. Stores that do provide unit pricing sometim play little tricks. They will show the unit price for two different bran but in different units. For example, they may show a national brand cereal in cents per ounce and a local brand as dollars per box. Or, o unit of measurement will be in ounces and a second in pounds. even worse situation arises when one product is measured in ounc and another in kilograms. How are you to know which is the bet buy? You need to convert one unit of measurement to another.

You might want to memorize or keep handy this list of conversio

Weight:	1 pound	= 16 ounces
	1 kilogram	= 1,000 grams
	1 kilogram	= 2.2 pounds
	1 pound	= 0.454 kilograms
Volume:	1 gallon	= 4 quarts = 128 ounces
	1 quart	= 2 pints = 32 ounces

1 pint   = 16 ounces
1 liter   = 1,000 millileters
1 liter   = 1.06 quarts
1 liter   = 34 ounces
1 ounce = 29.6 millileters

Suppose you have a half pint of jam at $3.52 from California and 155-milliliter jar of jam from France for $4.32. Which is the lower ice? Both jars are measured by volume. Convert the milliliters to inces: 155 milliliters is 0.155 liters, and 1 liter is 34 ounces. You can ultiply the 0.155 liters by 34 to find the volume of the French jar in inces.

Volume of French Jar = (0.155) × (34) = 5.27 ounces

Here's the unit price per ounce for the French jam.

Unit price = ($4.32)/(5.27 ounces) = $0.82 per ounce

One half pint is 8 ounces. Here's the unit price of the California m in ounces.

Unit price = ($3.52)/(8 ounces) = $0.44 per ounce

The California jam has the lower unit price. Try doing some "com-rison shopping" with the sample problems.

PROBLEMS

. You find two brands of potato chips. Brand X is 3.2 ounces for $1.69 and Brand Y is 8 ounces for $2.98. Which has the lower unit price?

. A 105-milliliter tin of imported caviar is $15.21. A 3-ounce jar of domestic caviar is $12.62. Which has the lower unit price?

. You are producing a local brand of clam chowder in a 12-ounce can. The popular national brand is selling for $1.39 for a 10-ounce can. You wish to undersell the competition by 10%. What should you charge for your can of clam chowder?

ANSWERS

1. This is a straightforward problem of computing the unit price for both brands of potato chips.

Unit price (Brand X) = ($1.69)/(3.2 ounces) = 52.8 cents/oz

Unit price (Brand Y) = ($2.98)/(8 ounces) = 37.2 cents/oz

Obviously, Brand Y is a cheaper brand.

2. This is basically the same as problem 1, but you need to convert the 105 milliliters into ounces. Since ounce = 29.6 milliliters, 105 milliliters = 3.55 ounces. The tin of caviar has the following unit price.

Unit price (tin) = ($15.21)/(3.55 ounces) = $4.28 per oz

The jar of caviar has the following unit price.

Unit price (jar) = ($12.62)/(3 ounces) = $4.21 per oz

The jar is slightly cheaper than the tin.

3. This problem is a little more complex. First you must compute the unit price of the competition.

Unit price (competition) = ($1.39)/(10 ounces) = $0.139 per oz

You want your chowder to be 10% less. Therefore, multiply the competition's unit price by 90% to your unit price.

Your unit price = (.90) × ($0.139) = $0.125 per oz

Now employ the unit price formula to compute the price of your 12-ounce can.

$0.125 = (Price)/(12 ounces)

Isolate the price and solve.

Price = ($0.125) × (12 ounces) = $1.50 per can

DISTANCE PROBLEMS

Distance problems have a notoriously bad reputation. My mother ha math phobia and would frequently say, "When I hear the words, 'Th train leaves the station at such-and-such time . . . ,' well, I know rig then that I'm lost." However, this reputation is unwarranted, becau distance and travel problems are no harder than the problems you'v been doing. Here's the basic formula for distance problems.

$$\text{Distance} = (\text{rate}) \times (\text{time})$$

The formula says that the distance you travel is equal to the rate which you are traveling multiplied by the length of time for which you travel. Look at an example. If you are driving at 55 miles per hour, and you drive for 4 hours, how far do you go? Enter the known values into the formula.

$$\text{Distance} \ = \ (\text{rate}) \ \times \ (\text{time}) \ = \ (55 \text{ miles/hour}) \ \times \ (4 \text{ hours})$$
$$= \ 220 \text{ miles}$$

Notice in the second step that the hours cancel out to leave the unit measure of miles, which is the unit of measurement you want. Sometimes you are asked to solve for either rate or time. If you know that you must drive 320 miles and that you will average 40 miles per hour, how long will it take?

$$320 \text{ miles} \ = \ (40 \text{ miles/hour}) \ \times \ (\text{time})$$

You need to manipulate the equation to solve for time. Here's a little reminder of how to isolate time on the left side of the equation. First divide both sides of the equation by the rate (40 miles/hour), which cancels the rate on the right side.

$$(320 \text{ miles})/(40 \text{ miles/hour}) \ = \ \text{time}$$

Now exchange the two sides of the equation.

$$\text{Time} \ = \ (320 \text{ miles})/(40 \text{ miles/hour}) = 8 \text{ hours}$$

You might want to see if you can travel on roads with speed limits over 40 miles per hour to shorten your trip!

You can also solve for rate. If you want to drive 450 miles in 9 hours, how fast must you drive? Once again, use the formula.

$$450 \text{ miles} \ = \ (\text{rate}) \ \times (9 \text{ hours})$$

$$\text{Rate} \ = \ (450 \text{ miles})/(9 \text{ hours}) = 50 \text{ miles/hour}$$

Distance problems can become quite complex, but the basic formula remains the same. Now try a few problems.

PROBLEMS

1. You're flying from Los Angeles, California to Memphis, Tennessee in a plane th͘
averages 350 miles per hour. If the distance is 1,877 miles, for how long will y͘
be flying?

2. The distance between New York City and Los Angeles, California is 2,911 mil͘
If you can walk it in 727.75 hours, how fast are you walking?

3. You take a bus to work. The bus travels at an average speed of 21 miles p͏
hour, and the trip takes 20 minutes. But after work, you miss the bus and ha͏
to walk home. Walking at 3.5 miles per hour, how long does it take?

ANSWERS

1. This is a straightforward problem to solve for time. Use the distance equation.

$$1,877 \text{ miles } = (350 \text{ miles/hour}) \times \text{ (time)}$$

Isolate time on the left and then solve for time.

$$\text{Time } = (1,877 \text{ miles})/(350 \text{ miles/hour}) = 5.36 \text{ hours}$$

2. This is a problem to solve for rate.

$$2,911 \text{ miles } = \text{ (rate) } \times (727.75 \text{ hours})$$

Isolate the rate and solve.

$$\text{Rate } = (2,911 \text{ miles})/(727.75 \text{ hours}) = 4 \text{ miles per hour}$$

3. This problem is a little more complex. From the first part of the problem you must determine how ma͏
miles it is to work. In order to do this, convert the 20 minutes to $\frac{1}{3}$ of an hour.

$$\text{Distance } = (21 \text{ miles/hour}) \times (\frac{1}{3} \text{ hour}) = 7 \text{ miles}$$

Now you can solve the second part of the problem.

$$7 \text{ miles } = (3.5 \text{ miles/hour}) \times \text{ (time)}$$

Isolate and solve for time.

$$\text{Time } = (7 \text{ miles})/(3.5 \text{ miles/hour}) = 2 \text{ hours}$$

WHAT HAVE YOU LEARNED?

You have reviewed a number of concepts and techniques in this cha͏
ter for solving real-world problems. First you explored unit of me͏
surement, and the procedure for manipulating units in an equati͏

st as you would numbers. Next you learned a rule for rounding
imbers.

Rule

If the left-most digit you are eliminating is 5 or greater, add 1 to the right-most digit being saved. If the left-most digit being eliminated is 4 or less, drop it or replace it with 0.

In the third section you explored using formulas. The next sections
ovided formulas for solving everyday problems. Here's a summary
some of the formulas.

Electrical usage:	Watts = (volts) × (amperes)
	Cost = (rate) × (usage)
Gas consumption:	Mileage = $\dfrac{\text{miles driven}}{\text{gallons used}}$
Unit pricing:	Unit price = (price)/(number of units)
Distance problems:	Distance = (rate) × (time)

Next up: Area and volume!

CHAPTER 17

COMPUTING AREAS
AND VOLUMES

Area is used to determine the surface space of a shape. If you need to carpet a room, for example, you need to know the area of the room to determine how much carpeting will cover the whole floor.

SQUARES AND RECTANGLES

Squares have the easiest areas to compute. A square is an enclosed figure with four equal sides (Figure 5a). When two straight lines meet they form a "vee" shape. This vee is an *angle*. We measure the size of angles in *degrees*. If a vertical line meets a horizontal line, the angle they make consists of 90 degrees and is called a *right angle*. Therefore a square has four right angles. The formula for computing the area of a square is simple.

$$\text{Area of square} = (\text{length of one side})^2$$

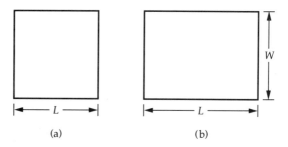

(a)   (b)

Figure 5. A square (a) and rectangle (b)

In formulas it is customary to use one letter of the alphabet to represent each value. Generally, you select the first letter of the first word. With this in mind, let "Area of square" = A, and "length of one side" = L. Here's the simplified formula.

$$A = L^2 \quad \text{(area of square)}$$

A rectangle is similar to a square, but one pair of opposite sides is longer than the other pair (Figure 5b). Like the square, the rectangle has four right angles. The formula for the area of a rectangle is the length times the width.

$$A = L \times W \quad \text{(area of rectangle)}$$

PROBLEMS

Compute each area.

1. Your house lot is a square, each side of which is 55 feet long. How many square feet are there in the lot?

2. Your porch is a rectangle with a length of 20 feet and a width of 16 feet. In square feet, what is the area of your porch?

ANSWERS

1. 3,025 square feet

2. 320 square feet

TRIANGLES

A triangle is a closed figure with three sides. The right triangle has one inside angle that is a right angle; it is the most common triangle you find in problems (Figure 6a). To compute the area of a right triangle, place one of the sides adjacent to the right angle horizontally so the other adjacent side is vertical. Measure the horizontal side and call

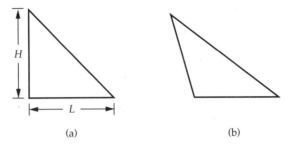

(a)   (b)

Figure 6. A right triangle (a) and oblique triangle (b)

it the length (L). Measure the vertical side and call it the triangle's height (H). Here is the formula for the area of a right triangle.

$$A = (^1/_2) \times L \times H \quad \text{(area of right triangle)}$$

If you look at Figure 7, you will recognize why the area of a triangle is just ½ the area of a rectangle with the same length and width (height). If you take a rectangle and divide it in half by drawing a diagonal line between opposite corners, you get two equal right triangles. The area of one right triangle is just half the area of the rectangle.

Figure 6 shows a triangle that is not a right triangle. Technically, such a triangle is called an oblique triangle. Computing the area of such a triangle is a little more difficult and requires a measurement on your part.

The first thing to do is to rotate the triangle until the longest side is horizontal. This side becomes the base. Once this is done, the opposite angle will be above the base side (Figure 8). Now drop a vertical line from the angle to the base side below it. You have changed the oblique

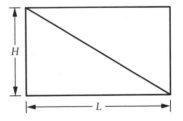

Figure 7. A rectangle divided into two right triangles

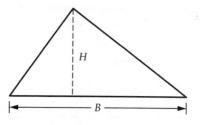

Figure 8. Computing the area of
an oblique triangle

riangle into two right triangles. Now you simply compute the areas of he two smaller triangles and add them. Let H stand for the height of he two triangles (the line you drew), and Y and Z stand for the two norizontal sides to the two triangles. Here's the formula for the area of the two triangles, and therefore, the area of the original triangle.

$$A = (1/2) \times H \times Y + (1/2) \times H \times Z$$

You can rewrite the equation.

$$A = (1/2) \times H \times (Y + Z)$$

But $Y + Z$ is just the length of the base, so you designate the length of the base as B. The formula becomes similar to the formula for the area of a right triangle.

$$A = (1/2) \times H \times B$$

Try the sample problems.

PROBLEMS

. If a right triangle has a length of 5.7 inches and a height of 4.3 inches, what is its area?

. An oblique triangle has been divided into two right triangles by dropping a line with a length of 7.5 inches (side H). If the base side is 9.3 inches, what is the total area of the oblique triangle?

ANSWERS

1. $A = (\frac{1}{2}) \times L \times H = (\frac{1}{2})(5.7)(4.3) = 12.255$ square inches

2. $A = (\frac{1}{2}) \times H \times B = (\frac{1}{2})(7.5)(9.3) = 34.875$ square inches

CIRCLES

Now you are ready to tackle problems with circles. You find the area of a circle with the following formula, in which A is the area, r is the radius, and pi (represented by the symbol π) is the ratio of the circle's diameter to the circle's circumference; π is represented approximately by the decimal 3.14.

$$A = \pi r^2$$

The radius of a circle is $\frac{1}{2}$ the diameter. Using the formula, compute the area of a circle with a radius of 2 feet.

$$A = \pi r^2 = (3.14) \times 2^2 = (3.14) \times 4 = 12.56 \text{ sq. ft.}$$

Now try the sample problems.

PROBLEMS

1. What is the area of a cross section of 1-inch pipe?

2. You have a gallon of paint to cover a circular dance floor that has a radius of 22 feet. The gallon of paint will cover 1,200 square feet. Do you have enough paint?

ANSWERS

1. 0.785 square inches

2. No. The dance floor is 1,520 square feet.

COMPLEX AREAS

n this section you will compute complex areas by separating them
nto the areas you have already studied: squares, rectangles, triangles,
ind circles. Consider the floor plan in Figure 9, a rectangular area with
wo modifications: an added half circle, area C, and a cut-off triangle,
irea R. You'll compute the area of the rectangle, add the area of the
ialf circle, and subtract the area of the triangle.

Area of floor = (area of rectangle) + ($^1/_2$) × (area of circle)

−(area of triangle)

First compute the area of the large rectangle.

Area of rectangle = $L \times W$ = 22 × 12 = 264 sq. ft.

Compute the area of the half circle by first computing the area of the
:ircle. Notice that the diameter of the half circle is shown to be 6 feet.
Divide this in half to get the circle's radius to use in the formula.

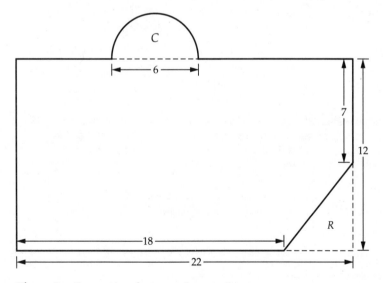

Figure 9.   Computing the area of a complex area

Area of circle $= \pi r^2 = (3.14) \times (3)^2 = (3.14) \times (9) = 28.26$

Divide the circle's area by 2 since we have only $1/2$ of the circle.

Area of half circle $= (28.26)/(2) = 14.13$ sq. ft.

Next compute the area of the triangle. Notice that Figure 9 does no show the lengths of the sides of the triangle. Yet, you can compute the missing lengths. Side L is just 22 feet minus 18 feet, or 4 feet. Side H is 12 feet minus 7 feet, or 5 feet. Now you can find the area of the triangle.

Area of triangle $= (1/2) \times L \times H = (1/2) \times (4) \times (5) = 10$ sq. ft.

Add the area of the half circle to the area of the large rectangle. From this value, subtract the area of the triangle. Your new value is equa to the area of the floor.

Area of floor $= 264 + 14.13 - 10 = 268.13$ sq. ft.

Although this example was long, the individual steps were simple and straightforward. The following sample problem is a bit more com plicated, but you should have no problem if you carefully write each step out.

PROBLEM

You are a real-estate agent who has just been granted exclusive rights to list the $6. million home of a movie star. You must, of course, compute the area of the variou rooms. Figure 10 is the floor plan of a front room. Based on the measurements giver compute the room's area.

ANSWER

This complex area has four different parts, the large rectangle, the small rectangle at the bottom righ the two half circles, and the alcove at the left. Computing the area of each and adding, you find a total flo space of 960.26 square feet.

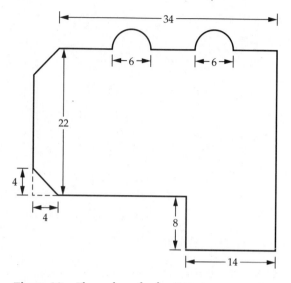

Figure 10.   Floor plan of a front room

CUBES AND RECTANGULAR VOLUMES

Now you'll progress from areas to volumes. Volume is how much three-dimensional space an object takes up. For example, if you were filling a fish tank, you would have to figure the volume of the tank to know how much water to put in. The simplest volume is that of a cube, which is an enclosed area with six square sides. The volume of a cube is given by the following formula, in which V is the volume, and L is the length of one side of one of the squares.

$$V = L^3$$

Using the formula, find the volume of a cube for which the length of one side measures 4.5 feet.

$$V = L^3 = (4.5)^3 = (4.5) \times (4.5) \times (4.5) = 91.125 \text{ cubic feet}$$

A rectangular volume is a volume enclosed by three pairs of rectangles (Figure 11). To compute the volume, simply multiply the lengths

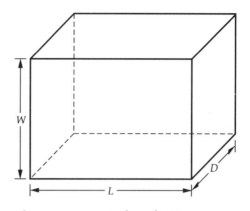

Figure 11.   Rectangular volume

of the sides of the different rectangles, which you will call the length, width, and depth (*L*, *W*, and *D*). Here's the formula.

$$\text{Volume} = L \times W \times D$$

A rectangular object has a length of 2.5 feet, a width of 1.4 feet, and a depth of 2 feet. Find its volume.

$$V = (2.5) \times (1.4) \times (2) = 7 \text{ cubic feet}$$

Now try the sample problems.

PROBLEMS

1. The average woman weighs 135 pounds. One cubic foot of water weighs 62 pounds. If the average female density approximates that of water, and the woman is melted down, how big a cubic box (in cubic feet) would be required to hold her? What would be the length of each side?

2. If you took all the wine the average Frenchman drank in 1981, it would fill a box 2.3 feet long, 1.3 feet wide, and 1 foot deep. If 1 cubic foot = 7.8 gallons, how many gallons of wine did the average Frenchman drink in 1981?

ANSWERS

1. Dividing 135 pounds by 62 pounds per cubic foot, you get 2.177. The box must contain 2.177 cubic feet, or 2.177 cu. ft. $= L^3$. Now simply solve for the length of one side, or L. Use the guessing method you learned in Chapter 15. Begin by guessing L to be 1.3 feet. Cubing L, you get 2.197, which is too big. By repeated tests, you find L to be approximately 1.296 feet. Hence, a cubic box with sides of 1.296 feet would approximate the volume of an average woman. (Although we don't recommend that you try melting down a 135-pound woman to check your answer!)

2. In this problem you must first calculate the volume of the box. Simply multiply the length by the width by the depth.

$$A = L \times W \times D = (2.3) \times (1.3) \times (1) = 2.99 \text{ cubic feet}$$

Now you simply multiply the cubic feet by gallons per cubic feet.

$$\text{Volume} = (2.99 \text{ cu. ft.}) \times (7.8 \text{ gallons/cu. ft.}) = 23.32 \text{ gallons}$$

Wow! Let's hope our average Frenchman spaced it out over the year and didn't drink all those gallons on New Year's Eve!

CYLINDERS AND SPHERES

A cylinder is simply a tube with a circle closing off each end (Figure 12). To compute the volume of a cylinder, you simply multiply the area of one of the circles by the length of the tube. Therefore, the

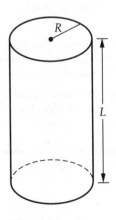

Figure 12. A cylinder

volume of a cylinder with a circle end that has radius r and length L can be found with this formula.

$$\text{Volume} = \pi r^2 L$$

Let's find the volume of a cylinder that has a radius of 2 inches and a length of 22 inches.

$$V = \pi r^2 L = (3.14) \times (2)^2 \times (22) = (3.14) \times (4) \times (22) = 276.32 \text{ in.}^3$$

The volume of a sphere is found with the following formula. A sphere is an object that is round all over—like a basketball.

$$\text{Volume} = (4/3)\pi r^3$$

Here you see another example of a number raised to its third power. What is the volume of a ball that is 1 foot in diameter? Remembering that the radius is $1/2$ of the diameter, you have the following solution.

$$\text{Volume} = (4/3)\pi r^3 = (4/3) \times (3.14) \times (0.5)^3 = 0.5233 \text{ ft}^3$$

You are such an expert now, that it's time for you to try the sample problems.

PROBLEMS

1. The average male brain occupies a volume of 88.5 cubic inches. A representative soda can is a cylinder with a diameter of 2.5 inches and a length of 4.5 inches. If the average man's brains were soda, how many cans would he be worth?

2. If you took everyone's eyes in the United States and compressed them into one giant eyeball, it would stand approximately 63 feet high. Assuming this eyeball were a sphere, how many cubic feet of eyeball would you have?

ANSWERS

1. For this problem you must compute the volume of the soda can and divide that volume into the brain's volume. Since the soda can is a cylinder, here's its volume.

$$V = \pi r^2 L = (3.14) \times (1.25)^2 \times (4.5) = 22.078 \text{ in.}^3$$

Divide this volume into 88.5 cubic inches to get 4 cans of soda.

2. For this problem, you have to calculate the volume of a sphere with a diameter of 63 feet, which divides to give us a radius of 31.5 feet.

$$V = (4/3)\pi r^3 = (4/3)(3.14)(31.5)^3 = 130,858 \text{ cubic feet}$$

Let's hope it's a good eyeball—it would be awfully hard to find a contact lens big enough to cover it.

WHAT HAVE YOU LEARNED?

In this chapter you have reviewed the use of formulas to compute areas and volumes. Consider keeping a copy of these formulas handy.

Area of a square $= L^2$

Area of a rectangle $= L \times W$

Area of a right triangle $= (1/2) \times L \times H$

Area of an oblique triangle $= (1/2) \times H \times B$

Area of a circle $= \pi r^2$

Volume of a cube $= L^3$

Rectangular volume $= L \times W \times D$

Volume of a cylinder $= \pi r^2 L$

Volume of a sphere $= (4/3)\pi r^3$

In addition, you have learned how to compute complex areas by using a combination of area formulas.

Stay tuned... interest rates are coming up.

CHAPTER 18

INTERESTING

INTEREST RATES

In this chapter you will delve into something truly interesting—interest. Car loans, credit cards, home mortgages, savings accounts, and investments all involve interest rate calculations, which, therefore, have a direct impact on our checkbooks and on our lives. Most of us, as consumers, do not often need to make such calculations on a daily basis. For the most part, banks, lending institutions, and brokerage firms do these computations for us. However, we all need to be "street smart" about how the rates are computed and about how we can manipulate interest to our best advantage.

SIMPLE INTEREST

In this section you'll explore how to calculate simple interest. If you borrow money, you must pay back the money borrowed plus interest. The money you originally borrowed is called the principal. The total you pay back is called the maturity value. Most of us want to know in advance the maturity value or the total we must pay back on a specific loan. The maturity value is dependent on the original principal, the interest rate, and the length of time for which you borrow the money. You can wrap this all up in a neat formula for simple interest.

$$S = P(1 + RT) \qquad \text{(simple interest)}$$

In the equation, S stands for the maturity value, P is the principal, R is the interest rate, and T is the length of time for which you borrowed the money. Suppose you borrow $1,000 from your friend Maria at 12% interest per year for 3 years. You can easily calculate what you must pay back. First change the interest rate from a percentage to a decimal; 12% = .12. Now substitute the values into the formula.

$$S = (1,000) \times (1 + .12 \times 3)$$
$$= (1,000) \times (1 + .36) = (1,000) \times (1.36) = \$1,360$$

You find out that you'll owe Maria a total of $1,360. Of the $1,360, the principal is $1,000, and the interest is $360. You can use this formula to find the interest simply by subtracting the principal from the answer.

When setting up such an equation, you must remember that the unit of measurement for the interest rate must be the same as that for the length of the loan. In the preceding example, the interest rate was 12% *per year,* and the length of the loan was 3 *years.* In both problems, you are dealing with years—the same unit of measurement. What if the interest rate had been 1% per month? You must now either change this rate into a yearly rate or change the length of the loan into months. Let's try the second approach. Three years is 36 months. Using an interest rate of 1% per month for 36 months, the formula looks like this.

$$S = (1,000) \times (1 + .01 \times 36) = (1,000) \times (1.36) = \$1,360$$

Notice that you get the same answer. And you should, because with simple interest, 1% per month is the same as 12% per year. Try the sample problems.

PROBLEMS

1. You agree to pay simple interest on $1,000 from your friend Sue for 5 years. How much simple interest will you pay if the rate is 5% per year?

2. You borrow $10,000 to buy a new car. You are to make one payment at the end of 3 years, with a simple interest rate of 12%. What is your entire payment, including both principal and interest?

ANSWERS

1. An interest rate of 5% and a term of 5 years produces the following:

$$S = (1,000) \times (1 + 0.05 \times 5) = (1,000)(1.25) = \$1,250$$

Since you borrowed $1,000, you will pay $250 interest.

2. For this problem, the formula yields this result.

$$S = (10,000) \times (1 + 0.12 \times 3) = (10,000) \times (1.36) = \$13,600$$

COMPOUND INTEREST

In the preceding problems you were concerned with only simple interest and with only one payment at the end of the loan. In the real world you are sometimes confronted with compounded interest or periodic payments. (These may complicate our lives, but since most of us are not able to buy a house or car in one lump payment, they're not such bad ideas.) While the formulas for computing compound interest and periodic payments require no more math than you have learned, they are complex. Therefore, consider the next three sections of this chapter to be optional. If you're still struggling with simple interest, you may want to skip to the section about credit cards and small loans. If you feel confident with your computing skills, come along and join the fray.

Compound interest works this way. You borrow money from the bank at a certain interest rate. After awhile, you owe the bank interest on that money. The bank adds this interest to the principal and, for the next time period, charges you interest on both the principal and the interest! This is a clever technique to get you to pay more interest than with simple interest.

Suppose, for example, you borrow $1,000 from the bank at 10% compound interest per year. The bank is going to compound the interest after 6 months. This means that the bank computes how much interest you owe for the first 6 months, and at the end of the year charges you interest on the principal plus the first 6 months' interest. You can compute the interest accrued for the first 6 months with the formula for simple interest.

$$S = (1,000) \times (1 + 0.10 \times 0.5) = (1,000) \times (1.05) = \$1,050$$

Since the original loan was for $1,000, the interest is $50. For the second 6 months, the bank is going to charge you interest on $1,050 instead of on $1,000. You can use the same formula to compute the total interest at the end of the second 6 months.

$$S = (1,050) \times (1 + 0.10 \times 0.5) = (1,050) \times (1.05) = \$1,102.50$$

If you had paid simple interest for the year at 10%, the total payment at the end of the year would have been $1,100, but because the bank compounded the interest after 6 months, the compounded maturity value is now $1,102.50. This is $2.50 more than with simple interest. Hence, the effective rate of interest was not 10% but 10.25%. This isn't much of an increase over the simple interest. However, as the interest rate goes up and the number of compoundings increases, the impact of compound interest becomes significantly greater.

Here's the innocent-looking formula for computing maturity value for compound interest rates.

$$S = P(1 + R)^N \qquad \text{(compound interest)}$$

In the above formula, N represents the number of times the interest is compounded (the number of compounding periods). The interest rate—R, in this case—must be the interest rate for the length of time of only one compounding. In the previous example, $N = 2$ because the interest was compounded every 6 months and there are 2 six-month periods in one year. Therefore, the interest rate, R, would be 5%, because the interest rate for an entire year is 10%, making the interest rate for 6 months one half that value. Let's plug these numbers into the new formula.

$$S = (1,000) \times (1 + 0.05)^2 = (1,000) \times (1.1025) = \$1,102.5$$

The maturity value that the compound-interest formula yields is exactly the same answer of $1,102.5. The new formula saved us a step.

Now let's apply the formula to a problem that's close to home. You borrow $10,000 to buy a new car, at 12% per year, with one payment at the end of 3 years. If the loan is compounded monthly, what will the final payment be? First compute R in the compound-interest formula. The compounding period is one month, and there are twelve months in a year. So the interest rate should be (12%)/12, or 1% per month. The total number of months (compounding periods) in the loan is 36 (3 years $\times$ 12 months/year). Thus, the formula looks like this.

$$S = (10,000) \times (1 + 0.01)^{36} = (10,000) \times (1.01)^{36}$$

How in blazes are you going to calculate $(1.01)^{36}$? To do so in the normal method requires multiplying (1.01) by itself 36 times. A monumental task. This is what makes the formula difficult to use. A number of ways exist to solve this tricky problem, including repetitive multiplying on a calculator, using compound-interest tables, and advanced mathematics. I'm going to let you off the hook for the example by simply telling you that the value of $(1.01)^{36}$ rounded to three places to the right of the decimal point is 1.431. If you plug this value into the formula, you get the following result.

$$S = (10,000) \times (1.431) = \$14,310$$

Look back at Problem 2 in the first section, and you will see that the maturity value for the same loan at simple interest was \$13,600. Changing from simple interest to interest compounded each month adds \$710 to the final payment. This increases the amount of interest you would have to pay by almost 20%. Try piquing your interest with the sample problems.

PROBLEMS

1. You borrow \$2,500 from a small loan company. The term of the loan is two years, the interest rate is 18% per year, and the interest is compounded every six months. If you must pay all of the maturity value back at the end of the loan, how much will you have to pay?

2. Your brother-in-law is going to lend you money against an inheritance of \$11,000 that you will receive in 5 years. He is going to charge you 12% interest, compounded yearly. What should the principal be so the one, final payment is \$11,000 in 5 years?

ANSWERS

1. This is a straightforward use of the compounding formula. The principal, P, is $2,500. The interest rate is 18% per year, or 9% per 6 months. Hence, $R = 9\%$. N in the formula is 4, since there are 4 six-month periods in two years.

$$S = (2,500) \times (1 + 0.09)^4 = (2,500) \times (1.4116) = \$3,529$$

2. Here's a slightly different problem. You still use the compound-interest formula, but you are asked to solve for the principal when you know the maturity value is $11,000. The interest rate, R, is 12%, and N is 5.

$$\$11,000 = P \times (1 + 0.12)^5$$

First manipulate the equation to isolate P on the left.

$$P = (11,000)/(1 + 0.12)^5$$

Now solve for P.

$$P = (11,000)/(1.12)^5 = (11,000)/(1.7623) = \$6,241.84$$

Therefore, your loan will be only $6,241.84 for a maturity value of $11,000. Your brother-in-law is not being very brotherly!

COMPUTING PAYMENTS

We have not yet examined the most common case, when you borrow money and then make periodic payments. The interest rate from one payment period to the next can be computed with either simple or compound interest. In this section you'll learn to compute periodic payments. You might ask why you can't simply take the maturity value for the end of the loan and divide by the number of payment periods. This wouldn't be fair, because when you make periodic payments, you are paying back some of the money as you go. Therefore, you do not get full use of the entire principal during the term of the loan.

The formula for computing periodic payments looks scary, but it's actually not any more difficult than the formula for computing compound interest. Here's the formula.

$$\text{Payment} = P\left[\frac{R}{1 - [1/(1 + R)^N]}\right]$$

You're already familiar with all the terms in the formula. P is the principal, R is the interest rate for one payment period, and N is the number of payment periods. When using the formula, you work in a piecemeal fashion. First you compute the $(1 + R)^N$ part. Then divide this into the number 1. Next subtract the result from 1. You then divide this result into R. The last step is to multiply by P. Think of Problem 1 from the previous section. You borrowed $2,500 at 18%, compounded semiannually for two years. In this example you are going to make payments every six months. Again, R is 9% per period, P is $2,500, and N is 4. Use the formula for calculating equal payments for each six-month period.

$$\text{Payment} = (2,500) \times \left[\frac{0.09}{1 - 1/(1 + .09)^4} \right]$$

Do the hard part first. Compute $(1 + .09)^4$, which turns out to be 1.4116. Plug it into the equation.

$$\text{Payment} = (2,500) \times \left[\frac{0.09}{1 - 1/1.4116} \right]$$

Now you divide 1.4116 into 1 to get 0.7084.

$$\text{Payment} = (2,500) \times \left[\frac{0.09}{1 - 0.7084} \right]$$

Subtract 0.7084 from 1 to yield 0.2916.

$$\text{Payment} = (2,500) \times \left[\frac{0.09}{0.2916} \right]$$

Divide 0.2916 into R, which is 0.09, to get 0.3086.

$$\text{Payment} = (2,500) \times (0.3086)$$

Now multiply 0.3086 by the principal for the final answer, Payment = $771.50.

Each payment is $771.50, and the total of all four payments is $3,086. You can see that the total paid with periodic payments is considerably less than the $3,529 which you would owe if the interest had compounded, and you waited until the end of the loan period to make one payment. Of course, the sum of the periodic payments should be less because you don't have use of the entire principal for the full term of the loan. Also, each period you pay the interest owing for that period plus an additional amount that is subtracted from the principal. Since the interest is paid off each period, it never has a chance to compound. The principal gets smaller rather than larger. Now try a problem on your own.

PROBLEM

You go to Chuck's Slightly Used Cars to buy a desperately needed automobile. Chuck sells you a nice car for $4,000 at 22% interest per year, to be paid in 4 quarterly payments over one year. He tells you the payments are $1,234.40 per quarter. Is Chuck cheating you? If you think so, what should each quarterly payment be?

ANSWER

The first thing to do, of course, is to compute the quarterly payments based on the information you have. The principal, P, is $4,000. The number of payment periods is 4. The yearly interest rate is 22%. This gives a quarterly interest rate of 0.22/4, or 0.055 per quarter. Now plug these values into the payment formula.

$$\text{Quarterly Payment} = (4,000) \times \left[\frac{0.055}{1 - 1/(1 + 0.055)^4} \right]$$

First compute $(1 + 0.055)^4$, which turns out to be 1.2388. Divide this number into 1 to get 0.8072. This number is subtracted from 1 to yield 0.1928. Divide 0.1928 into 0.055 to get 0.2853. The last step is to multiply 0.2853 by 4,000 to get $1,141.20. Hmmm. This is less than what Chuck said. Good old Chuck is overcharging you by $93.20 per quarter, for a total of $372.80. For him to charge you $1,234.40 per quarter, his yearly interest rate would have to be 36%.

USING INTEREST RATE TABLES

In the equations for computing compound interest and periodic payments, you encountered a difficult operation, the computation of $(1 + R)^N$. Whenever I want to calculate this value, I use a calculator and perform repetitive multiplications. Yet, there is a simpler way. Some books on investing and money management contain tables that calculate the value of $(1 + R)^N$ for you. Look at Table 18.1. This table shows N down the left side and R across the top. At the intersection of a particular R and N is the value for $(1 + R)^N$.

Table 18.1 Value of $(1 + R)^N$ for Various *N* and *R*

N	\multicolumn{6}{c}{R}
	0.50%	0.75%	1.00%	1.25%	1.50%	2.00%
1	1.0050	1.0075	1.0100	1.0125	1.0150	1.0200
2	1.0100	1.0151	1.0201	1.0252	1.0302	1.0404
3	1.0151	1.0227	1.0303	1.0380	1.0457	1.0612
4	1.0202	1.0303	1.0406	1.0509	1.0614	1.0824
5	1.0253	1.0381	1.0510	1.0641	1.0773	1.1041
6	1.0304	1.0459	1.0615	1.0774	1.0934	1.1262
7	1.0355	1.0537	1.0721	1.0909	1.1098	1.1487
8	1.0407	1.0616	1.0829	1.1045	1.1265	1.1717
9	1.0459	1.0696	1.0937	1.1183	1.1434	1.1951
10	1.0511	1.0776	1.1046	1.1323	1.1605	1.2190
11	1.0564	1.0857	1.1157	1.1464	1.1779	1.2434
12	1.0617	1.0938	1.1268	1.1608	1.1956	1.2682

If you want to know the value of $(1 + R)^N$ for a specific *N* and a specific *R*, you can just look it up in the table. If you are interested in $(1 + 0.01)^9$, look under the 1.00% column down to the row where *N* is 9. The value of $(1 + 0.01)^9$ is 1.0937. Bear in mind that the interest rates shown are for each compounding, or payment period. Thus, if the 1% is for monthly compounding, the yearly interest rate would be 12%. Tables in financial books generally contain many more values of *R* and *N*.

CREDIT CARDS AND SMALL LOANS

So far we have confined our exploration to the technical aspects of how interest and payments are computed. Now it's time to consider how interest affects our lives. In this section we'll investigate credit cards and small loans. In the next section we'll cover home mortgages.

Many of us use credit cards on a continuing basis and periodically take out small loans to cover the purchase of cars, larger household items, and vacations. If you pay your credit-card accounts off completely each month and secure low-interest small loans, you are using your borrowing power wisely, since credit cards usually have higher interest rates than do bank loans. Yet many of us often carry a continuing debt on our credit cards and are careless with small loans.

Let's consider credit cards first. If you carry a continuing debt on your credit cards of $10,000 (not an unusual situation), you are probably spending between $1,000 and $2,000 per year on interest. Since you will eventually have to pay for all the charged items anyway, this interest is a pure loss to you. If you itemize your income taxes and want to know the interest you are losing each year to credit cards, simply check your tax return.

At one time, you received a tax advantage with revolving credit because you could deduct the interest from your income and not pay taxes on it. No longer. During 1989, the IRS allowed you to deduct only 20% of this interest from your income. In the future this deduction will be completely lost. Such interest is an unnecessary drain on your income. Suppose you pay an average of $300 on your cards each month. Perhaps only $175 is really paying for goods—the other $125 pays some of your constantly accumulating interest. By paying off your cards each month, you can significantly reduce your monthly payments and increase the value of goods you are buying, thus improving your standard of living.

Small loans are another area in which people often pay more than necessary. Most of us have several possible sources of credit: banks, home equity, credit unions, and small lending companies. Before you purchase that car or borrow for that vacation, shop around for the lowest-interest-rate loan. A small loan with an interest rate from 9% to 12% will save many dollars compared to a loan with interest of 15% to 24%. Now that you know how interest is calculated, you can compare loans not only by the interest charged but also by whether or not the interest is compounded and how often the compounding is done. You can also benefit from loans for which early repayment is not penalized.

HOME MORTGAGES

Home mortgages are important loans because they generally involve large amounts of money borrowed over many years. A typical home mortgage can secure $100,000 borrowed over 30 years. Yearly interest rates vary greatly. As I write this, average mortgage interest rates are approximately 10.5% per year. At that rate of interest, a $100,000 mortgage will have a monthly payment of $914.74, which amounts to total payment of $329,306.40 over 30 years. The $914.74 monthly payment includes money paid for interest on the loan with a fraction going to pay off the principal. The payment does not include taxes or insurance. Actual monthly payments on a home can be greater if these two expenses are added to the mortgage payment.

The monthly payment is computed with the periodic payment formula. If the yearly interest rate is 10.5%, then the monthly rate is 0.105/12, or 0.00875. You can use an amortization table to track mortgages. Table 18.2 shows the amortization table for the first twelve months of the example loan. Amortization tables for various loans at different amounts are available from office-supply stores.

Notice from Table 18.2 that each payment is divided between interest and the amount being subtracted from the principal. For the first month, $875 is interest. This is just as it should be, since the original loan is $100,000 and the monthly interest rate is 0.00875; $100,000 \times 0.00875 = \$875$. This is all interest. Subtract the balance of the payment, $39.74, from $100,000 to get $99,960.26. For the second month, the interest rate of 0.00875 is multiplied by what you still owe, $99,960.26, to get the interest for the second month, or $874.65. Subtract this amount from the total payment to get $40.09. Then subtract $40.09 from the loan balance to get the new balance. This process continues for 30 years, or 360 months, until the loan is paid off. You can see why such a loan takes 30 years to pay off. During the first year, you pay off only $500 of the original $100,000.

Table 18.2 Amortization of $100,000 Mortgage at 10.5% Over 30 Years

Monthly Payment Number	Amount Going to Principal	Amount Paid for Interest	Remaining Balance of Loan
1	39.74	875.00	99,960.26
2	40.09	874.65	99,920.16
3	40.44	874.30	99,879.72
4	40.79	873.95	99,838.94
5	41.15	873.59	99,797.79
6	41.51	873.23	99,756.28
7	41.87	872.87	99,714.41
8	42.24	872.50	99,672.18
9	42.61	872.13	99,629.56
10	42.98	871.76	99,586.59
11	43.36	871.38	99,543.22
12	43.74	871.00	99,499.49

For the home buyer there are two major considerations: the basic loan you get to begin with and the way you are going to pay off that loan. Always remember the interest rate. A small change in the interest rate can involve a large change in the overall loan repayment. Suppose you secured the $100,000 loan at 9.5% instead of 10.5%. Your monthly payments would drop to $840.85. This is a yearly savings of $886.68 and a total savings over the life of the loan of $26,600.40. On the other hand, suppose you had to pay 12% yearly interest. Here, the monthly payments would increase to $1,028.61 for a yearly increase of $1,366.44. Over 30 years, this would cost an extra $40,993.20.

Looking at these examples, you can appreciate how important your mortgage interest rate is. When buying a home, always shop for the lowest rates.

Most people believe that once their mortgage is signed, they are locked into the terms of the loan until the loan is paid off, or until they sell the house and retire the loan. This is not always the case. Some mortgage contracts contain a penalty clause for early payment of the principal. This clause is included to discourage you from paying the loan off earlier than the amortization table specifies. Many mortgages, however, contain no such penalty clause. In fact, some monthly mortgage payment forms contain a place to write in any extra principal payment you want to make. All of this extra payment is subtracted directly from your balance, reducing future interest and increasing the future amounts going toward paying the balance off. This has the effect of reducing the number of years of your mortgage contract.

Is it wise to make extra payments beyond the specified mortgage payment? In many cases, it is.

To see how increasing your monthly mortgage payment can benefit you, consider a loan of $100,000 for 20 years instead of 30 years. The monthly payment on the 20-year loan is $998.38 per month, or $239,611.20 over the 20 years. This 20-year loan is only $83.64 a month more. Over the course of the loan, you save $89,695.20 and pay off your loan 10 years sooner.

Why wouldn't you get a 20-year loan rather than a 30-year loan? Often, the size of the loan is so great that the 20-year payment schedule is frightening or threatens to have too big an impact on your budget. Accordingly, you choose the smaller monthly payment of the 30-year loan. However, even if you have chosen the 30-year loan, you can still turn it into a 20-year loan by paying a little extra each month.

First check to see whether your present mortgage has a penalty clause. If not, decide on the number of years you want to take to pay off your mortgage, and check your amortization schedule (available from your mortgage company or local office-supply store) to see how much you still owe. Then ask your mortgage company or your bank to compute the monthly payments on a mortgage equal to the amount you still owe, for the number of years you've selected, at your current interest rate. When they give you that figure, increase your monthly payments enough to match that mortgage. I recommend that you not refinance, since this is usually an expensive proposition. Simply increase your current payments.

Let's take an example. Your current mortgage contract is for $80,000 over 30 years at 10.5% interest. You have been paying on this mortgage for 3 years and still have 27 years to go. You decide you want it paid off 20 years from now. Your current monthly mortgage payment is $731.79 (excluding taxes and insurance), and you still owe $78,662 of the original $80,000.

The bank tells you a mortgage for $78,662 for 20 years at 10.5% interest would have a monthly payment of $785.35. This is only $53.56 more than your current monthly payment, but your mortgage will be paid off 7 years early, with a total savings of $48,615.96 in interest over the life of the mortgage.

The nice feature of this overpayment is that, if you have money problems, you can stop the extra payment until you have recovered. The choice of extra payment is always up to you. The more you overpay, the shorter the loan, and the more money you save in the long run.

WHAT HAVE YOU LEARNED?

You have learned three handy formulas for computing simple interest, compound interest, and periodic payments.

$$S = P(1 + RT)$$ (Simple Interest)

$$S = P(1 + R)^N$$ (Compound Interest)

$$\text{Payment} = P\left[\frac{R}{1 - 1/(1 + R)^N}\right]$$ (Periodic Payments)

You have also seen how tables can be used to find $(1 + R)^N$.

In addition to the material on computing rates and payments, you have read some general comments about using credit. Credit cards are most helpful when paid off each month, and small loans are most wisely secured by shopping for the best interest rates. Of course, you should shop for the best mortgage when buying a home. If you are already participating in a mortgage contract, consider overpayment of the monthly mortgage to reduce the term of the loan and save substantial money over the life of the contract.

After all this, you deserve some basic fun. So flip the page and find out how to do math in your head. Look, Uncle Ozzie, no hands!

CHAPTER 19

THE JOYS OF MENTAL
ARITHMETIC

When you actually need to do arithmetic, are you always sitting comfortably at your office desk or your dining-room table, with easy access to paper, pencil, and calculator? Usually not. More often, you need arithmetic at odd, unpredictable times when you are away from paper, pencil, and calculator—when you are engaged in other activities. You might be driving your car, shopping, mowing the lawn, preparing dinner, or standing at a pay phone. Suddenly, you need an answer. What's the magic solution? Mental arithmetic.

Most schools spend little time on mental arithmetic, but much of the math you need as an adult involves only mental computation. You already know some mental arithmetic. If you have used 6 eggs in a recipe that calls for 8, you subtract the 6 from the 8 and realize you need 2 more eggs. This level of computing seems almost intuitive; you forget that you are doing arithmetic. In this chapter you'll meet some techniques that can expand your calculating powers.

ADDING ON

The first technique is one you already know—adding a small number to a larger number. You've carried in 3 logs for the fireplace and you carry in 2 more. You mentally add the 2 logs and know you have 5 logs. Suppose you carry in 13 logs and then carry in 2 more. You add 2 and 13 in your head to get 15. Why does this work? It works because of the place-value system. In each of the following examples, the only digit that changes is the ones digit.

$$3 + 2 = 5$$

$$13 + 2 = 15$$

$$23 + 2 = 25$$

$$33 + 2 = 35$$

Adding 2 to any number is a kind of extended counting. You normally count by adding 1 to the preceding number. To add by 2s just add an additional 1. You can practice this by simply adding by twos: 2, 4, 6, 8 . . . or 1, 3, 5, 7 . . . . In fact, you should already be able to count from either 0 or 1 by twos to 100 without any trouble. This technique is also called skip-counting.

To extend this, you should, with ease, be able to add the number 2 to any number whose digits you can conveniently remember. When adding 2 to a number ending in 8 or 9, it is necessary to carry a 1 to the tens place. Larger numbers are a bit harder to do because it's harder to remember more digits.

$$4,388 + 2 = 4,390$$

$$17,619 + 2 = 17,621$$

Most of us can add or skip-count by twos. Let's expand this notion by adding with threes. As you try it, if you want to, use finger-counting—go ahead. There's nothing wrong with counting on your fingers, as you know, but the goal is to become so familiar with adding small numbers that you can do it automatically.

You can see where this leads. The goal is to be able to add any number from 0 to 9 to any other number whose digits you can conveniently remember. Suppose you have invited 38 children to your daughter's birthday party. She wants to invite 7 more children. How many party treats must you buy? You know that 7 plus 8 is 15. Since 7 plus 8 ends with a 5, then 7 plus 38 must be 45. That's a lot of party treats, but it is, after all, for your daughter's birthday. You can practice this mental arithmetic anywhere, at any time. If you are cooking vegetables in the microwave for 8 minutes, and the digital clock reads

9:44, try mentally adding the 8 minutes; you get 9:52. Set the timer and cook the vegetables. When the microwave's timer bell sounds, what time does the clock read? (If it's not 9:52, your timer needs repair.)

PROBLEMS

Carry out each addition in your head.

1. $13 + 5 = ?$

2. $27 + 2 = ?$

3. $63 + 4 = ?$

4. $49 + 7 = ?$

5. $89 + 8 = ?$

6. $101 + 4 = ?$

7. $273 + 6 = ?$

8. $436 + 9 = ?$

9. $1,934 + 4 = ?$

10. $5,338 + 7 = ?$

ANSWERS

1. 18

2. 29

3. 67

4. 56

5. 97

6. 105

7. 279

8. 445

9. 1,938

10. 5,345

ADDING BY TENS

You can easily extend the technique of adding on to adding larger numbers that end in 0. The numbers between 10 and 100 should be no problem.

$$10 + 30 = 40$$

$$60 + 20 = 80$$

$$30 + 50 = 80$$

$$70 + 40 = 110$$

In the preceding examples, you are just adding two single-digit numbers and leaving the zeros unchanged. You can get away with this because the zeros have no value; they are just placeholders. Now try hundreds. You apply the same principle.

$$100 + 300 = 400$$

$$600 + 200 = 800$$

$$300 + 500 = 800$$

$$700 + 400 = 1,100$$

When adding 700 and 400, remember 7 and 4. Realizing that 7 + 4 = 11, replace the two zeros to get 1,100.

You can mentally add any two numbers that have all zeros except the left-most digit. You can add hundreds, thousands, tens of thousands, even millions: 5 million plus 6 million is 11 million; 30 million plus 20 million is 50 million.

You can also mix numbers of different places and easily add them in your head; 100 plus 20 is 120. You don't have to add 2 to 10 to get 12; you can simply fill in the middle digit of the 100 with the 2 of the 20. Look at some examples.

$$300 + 40 = 340$$

$$1100 + 6 = 1106$$

$$9 + 10,000 = 10,009$$

$$1300 + 10 = 1310$$

In each case you don't really have to add, because none of the digits change. You just combine numbers. Try some more mental addition.

PROBLEMS

Carry out each addition in your head.

1. 10 + 30 = ?

2. 60 + 70 = ?

3. 30 + 100 = ?

4. 120 + 90 = ?

5. 600 + 40 = ?

6. 500 + 400 = ?

7. 700 + 600 = ?

8. 1,100 + 20 = ?

9. 1,000 + 200 = ?

10. 10,400 + 600 = ?

ANSWERS

1. 40

2. 130

3. 130

4. 210

5. 640

6. 900

7. 1,300

8. 1,120

9. 1,200

10. 11,000

ADDING LARGE NUMBERS

Now the fun begins! Using the techniques of adding on and adding numbers ending in zeros, you can add large numbers in your head. Let's start with numbers between 10 and 100. To add the numbers 42 and 78, you perform a little dissection. Separate 42 into 40 plus 2, and 78 into 70 plus 8.

$$42 = 40 + 2$$

$$78 = 70 + 8$$

Add 40 and 70 to get 110, and 2 and 8 to get 10.

$$40 + 70 = 110$$

$$2 + 8 = 10$$

Now add 110 and 10 to get 120.

$$110 + 10 = 120$$

It's that simple. Just separate the numbers into component parts, add the various parts, and then reassemble them. This works easily even with larger numbers.

Let's add 491 to 365. First separate the parts.

$$491 = 400 + 90 + 1$$

$$365 = 300 + 60 + 5$$

Add hundreds, then tens, and then ones.

$$400 + 300 = 700$$

$$90 + 60 = 150$$

$$1 + 5 = 6$$

Now reassemble the parts.

$$700 + 150 = 850$$

$$850 + 6 = 856$$

The answer is 856. The trick is to memorize the two original numbers so you can keep referring to them for the next set of digits. After memorizing the original numbers, you begin with the large digits (the digits in the greatest place) and add them. Store this answer in your memory. Then you call up the original numbers again and move one place (or digit) to the right. So, in the preceding example, you began by adding the digits in the hundreds place, stored the answer, and moved to the tens place. After adding the sum of the digits in the tens place to the 700 in memory, you stored the 850, and then, remembering that the original numbers were 491 and 365, moved to the ones place and added 1 and 5.

You don't have to visualize separating the 491 into 400 + 91; simply visualize the 4 in the hundreds place. Take the 3 from 364, and add it to 4 to get 7. Knowing you were working with the hundreds place, you store 700 into memory. Not having to visualize 400 + 91 saves a step. Examine the following rule.

Rule
When mentally adding two large numbers, separate and add individual digits, beginning at the left, moving to the right, and ending with the ones digits.

Look at some examples.

(the problem)	211 + 489
(adding hundreds)	2 + 4 = 600; store 600
(adding tens)	1 + 8 = 90
	600 + 90 = 690; store 690
(adding ones)	1 + 9 = 10
(the answer)	690 + 10 = 700

(the problem)	674 + 441
(adding hundreds)	6 + 4 = 10 hundreds, or 1,000; store 1,000
(adding tens)	7 + 4 = 110
	1,000 + 110 = 1,110; store 1,110
(adding ones)	4 + 1 = 5
(the answer)	1,110 + 5 = 1,115

This technique works well with money, too. When adding money, add the dollars first. Then add the cents and, in the final step, add the cents to the dollars.

(the problem)	$11.42 + $26.31
(adding tens)	1 + 2 = $30; store $30
(adding ones)	1 + 6 = $7
(adding ones and tens)	30 + 7 = $37
(adding cents)	42 + 31 = 73 cents
(adding cents to dollars)	$37 + 73 cents = $37.73

In the money example, you can take a shortcut if you are confident enough about your skills. In the second step you can add 11 to 26 directly to get 37. Then you add the cents to get $37.73.

(the problem)	$11.42 + $26.31
(adding tens and ones)	11 + 26 = $37; store $37
(adding cents)	42 + 31 = 73 cents
(adding cents to dollars)	$37 + 73 cents = $37.73

Try the sample problems.

PROBLEMS

Mentally add each pair of numbers.

1. 21 + 40 = ?

2. 87 + 62 = ?

3. 59 + 44 = ?

4. 110 + 73 = ?

5. 143 + 216 = ?

6. 439 + 328 = ?

7. 691 + 922 = ?

8. 789 + 184 = ?

9. 338 + 654 = ?

10. 1,422 + 368 = ?

ANSWERS

1. 21 + 40: 2 + 4 = 60; 60 + 1 = 61

2. 87 + 62: 8 + 6 = 140; 7 + 2 = 9; 140 + 9 = 149

3. 59 + 44: 5 + 4 = 90; 9 + 4 = 13; 90 + 13 = 103

4. 110 + 73: 11 + 7 = 180; 180 + 3 = 183

5. 143 + 216: 1 + 2 = 300; 4 + 1 = 50; 300 + 50 = 350; 3 + 6 = 9; 350 + 9 = 359

6. 439 + 328: 4 + 3 = 700; 3 + 2 = 50; 700 + 50 = 750; 9 + 8 = 17: 750 + 17 = 767

7. 691 + 922: 6 + 9 = 1500; 9 + 2 = 110; 1500 + 110 = 1610; 1 + 2 = 3; 1610 + 3 = 1,613

8. 789 + 184: 7 + 1 = 800; 8 + 8 = 160; 800 + 160 = 960; 9 + 4 = 13; 960 + 13 = 973

9. 338 + 654: 3 + 6 = 900; 3 + 5 = 80; 900 + 80 = 980; 8 + 4 = 12; 980 + 12 = 992

10. 1,422 + 368: 14 + 3 = 1700; 2 + 6 = 80; 1700 + 80 = 1780; 2 + 8 = 10; 1780 + 10 = 1790

SUBTRACTION

Can you subtract large numbers? Yes. But mental subtraction of large numbers is more difficult than mental addition. Always subtract the smaller number from the larger number. Separate the digits of the smaller number in sequence and subtract them from the larger number. When finished determine whether the answer is positive or negative: if the smaller number was to be subtracted from the larger number in the original problem, then the answer is positive; if the larger number was being subtracted from the smaller number in the original problem, then the answer is negative. Examine the rule.

Rule

To mentally subtract with large numbers, identify the smaller number, begin with the digits at the left, and subtract these digits from the larger number. When finished, determine whether the answer should be positive or negative.

Let's take the example 493 minus 361. Begin with the 3 in the hundreds place of 361. Subtract it from the 4 in the hundreds place of 493 to get 193. Now, remembering the 193, subtract the 6 in 361 from the 9 in the tens place of 193 to get 133. Now subtract the 1 in 361 from the ones place of 133 to get 132. Look at the steps.

(the problem)	493 − 361
(subtracting the hundreds)	4 − 3 gives 193
(subtracting the tens)	9 − 6 gives 133
(subtracting the ones)	3 − 1 gives 132

Seems easy so far, but tiny traps lie in wait. Consider 912 minus 697. Subtract 6 from 9 to get 312. Now subtract 9 from the 1 in 312. But that won't work; so you either have to realize that 31 minus 9 is 22, or you have to mentally carry 1 from the 3 and make the 1 an 11 before subtracting—that could get messy. If you know that 9 from 31 is 22, the 312 becomes 222. Now subtract 7 from 2. Again, messiness. If you know that 7 from 22 is 15, you're okay. The answer is 215.

However, in subtracting 697 from 912, you could have used a little trick to find the answer. Since 697 is very close to 700, just subtract

⁷00 from 912 to get 212. Remember that you subtracted 3 too many, and just add the 3 back: 212 plus 3 is 215. This method is much ɛasier; it works well because it turns a process of subtracting several ɩimes into one of subtracting once and then adding, and adding is an ɛasier operation.

When you practice addition and subtraction of long numbers, you ɩan cheat a little. Write the two numbers in the problem on a piece ɔf paper. This will relieve yóu of remembering the original problem vhile you do your mental computations. Look at the two numbers ɩnvolved, and imagine the manipulations required to get the answer. ɩhen write the answer down and check it. Once you've mastered this, ʾou can abandon writing any numbers down and carry out the entire ɔrocess in your head. Try the sample problems.

PROBLEMS

Ɂarry out each subtraction in your head. The method you choose may not be identical ɔ that shown in the answers.

1. 47 − 32 = ?

2. 69 − 15 = ?

3. 43 − 18 = ?

4. 61 − 35 = ?

5. 155 − 32 = ?

6. 568 − 133 = ?

7. 719 − 304 = ?

8. 815 − 481 = ?

9. 511 − 287 = ?

10. 1,209 − 491 = ?

ANSWERS

1. 47 − 32: 47 − 30 = 17; 17 − 2 = 15

2. 69 − 15: 69 − 10 = 59; 59 − 5 = 54

3. 43 − 18: 43 − 10 = 33; 33 − 8 = 25

4. 61 − 35: 61 − 30 = 31; 31 − 5 = 26

5. 155 − 32: 155 − 30 = 125; 125 − 2 = 123

6. 568 − 133: 568 − 100 = 468; 468 − 30 = 438; 438 − 3 = 435

7. 719 − 304: 719 − 300 = 419; 419 − 4 = 415

8. 815 − 481: 815 − 400 = 415; 415 − 80 = 335; 335 − 1 = 334
or
815 − 500 = 315; 315 + 19 = 334

9. 511 − 287: 511 − 200 = 311; 311 − 80 = 231; 231 − 7 = 224
or
511 − 300 = 211; 211 + 13 = 224

10. 1,209 − 491: 1,209 − 400 = 809; 809 − 90 = 719; 719 − 1 = 718
or
1,209 − 500 = 709; 709 + 9 = 718

APPROXIMATING

Before tackling mental multiplication, let's consider the technique of approximating the answer to a problem rather than going for the exact answer. Problems often don't require an exact answer. To figure out how much gas you need to get to the next town, do you need the answer down to ounces or fractions of an ounce? If you're cooking for a large group, do you care whether you're increasing the recipe amounts enough to serve *exactly* 17 people?

Being satisfied with approximate answers opens a completely new field of computation, because you can mentally solve many more kinds of problems. The process of approximation is easy. Just carry out the operations required for the greater place values, ignoring the lesser place values. This is the reverse of the normal procedure of working with numbers from right to left. Computing with the greater places first means moving from the left to right. Suppose you're at the local grange, getting weekend supplies. The clerk checking out your groceries seems absentminded. The total on the receipt is $47.51. Is this number "in the ballpark"?

Let's check the total by adding up the approximate prices for each item. You bought a gallon of antifreeze for the car at $9.99. No problem; you'll just pretend that it was $10. You also bought a 50-pound sack of duck feed for $8.65. Call that $9 and add it to the $10; that's $19. At this point, you know you're a little over, but by less than a dollar. Besides, taxes will help absorb some of the error. You also put $15.21 worth of gas in the car. Rounding down to $15, add $15 to $19 to get $34 (15 plus 20 is 35; then subtract 1). A new garden hose for $7.95 (the dog ate the last one) makes this $8 added to $34 or $42. You must be off somewhere, because the final bill is $47.51; that's a difference of $5.51. Oh, yes, you bought some batteries for $3.48. That's close to $3.50, and $3.50 plus $42 is $45.50. The receipt must be reasonably correct. If you made the effort to add the exact amounts, you would have come up with $45.28—a difference of 22 cents out of 45 dollars. The difference between $45.28 and $47.51 is accounted for by sales tax. Absentminded or not, the clerk rang up the correct amount.

There is a good reason to learn this system. Modern cash registers seldom make adding mistakes. The mistakes are usually made by clerks who enter the wrong price, enter duplicate items, or forget items on sale. You can catch this kind of mistake by mentally approximating the cost of your purchases as you wait in the checkout line. Then when the total appears, you know whether a serious error has occurred. Don't be embarrassed about using this technique to question the total. The cashier can quickly add the items again. After all, it's your money that is at stake, and you are the customer.

PROBLEMS

Add each list of numbers in your head by rounding to the nearest dollar.

1. $1.22 + 3.67 + 1.89 + 5.59 = ?

2. $0.22 + 8.33 + 2.99 + 7.16 + 3.49 = ?

3. $3.89 + 0.89 + 11.12 + 7.89 + 3.11 + 0.77 = ?

ANSWERS

The exact answer is given in parentheses.

1. $13 ($12.37)   **2.** $21 ($22.19)   **3.** $28 ($27.67)

MULTIPLICATION

Let's apply the approximation technique to mental multiplication. Begin by multiplying the greater places, and then decide if you need a more accurate answer. If you do, continue multiplying the lower places; if you are satisfied with the answer, just stop.

As with addition, you break the multiplication problem into part and work on those parts. The key is to recognize the best way to break down the numbers. This takes practice and becomes almost an art. Let's begin with 50 × 33. Here, you focus on that 50. Multiplying by 50 is easy, because you can simply multiply by 100 and then cut the answer in half. So, 100 times 33 is 3,300. Divide 3,300 in half You know that half of 32 is 16. Therefore, half of 3,200 is 1,600. You must be close: 3,300 is just 100 greater than 3,200. Divide this 100 by 2 to get 50, and add 50 to 1,600 to get 1,650. Think of it this way 100 times 33 is 3,300, and half of 3,300 is 1,650.

Try a slightly harder problem: 53 × 33. You know already that 50 × 33 = 1,650, but you want to solve 53 × 33. The approximation 1,650 is just a little off—in fact, just 3 × 33 off—because 53 × 33 = (50 + 3) × 33, which is equal to 50 × 33 + 3 × 33. You've already done 50 × 33; now add 3 x 33. which is 99. Add 99 to 1,650 to get 1,749 (that's 1,650 + 100; then subtract 1). What starts out looking like a tough multiplication problem reduces to a series of simple steps

To sum up mental multiplication: First decide that you are going to estimate your answer to the degree of accuracy you need. Then think about the two numbers being multiplied. Select the one that can be conveniently rounded to a number ending in 0 or can be easily broken down into numbers ending in 0. Complete the multiplication with the rounded numbers. For more accuracy, estimate how much should be added back into your answer or subtracted from it, and carry out the necessary process.

Rule

To mentally multiply two large numbers, study the numbers and then round or break them into parts to carry out easy multiplication. For more accuracy, adjust the answer by adding or subtracting portions missed due to rounding.

PROBLEMS

Mentally perform each multiplication.

1. 10 × 44

2. 12 × 48

3. 150 × 32

4. 120 × 24

5. 340 × 71

ANSWERS

Your answers may not show as much accuracy as the printed answers.

1. Shift 44 left one place, and add a 0. 10 × 44 = 440

2. 12 × 48: Notice that 10 × 50 = 500. This is a first approximation. Next get a closer estimate, 12 × 50. You know that 2 × 50 = 100. Since 12 × 50 = 10 × 50 + 2 × 50, you can add 100 to 500 to get 12 × 50 = 600. You can get the exact answer by realizing that 12 × 48 = 12 × 50 − 12 × 2. Since 12 × 2 = 24, you have 12 × 48 = 12 × 50 − 12 × 2 = 600 − 24 = 576.

3. 150 × 32: First, 150 × 10 = 1,500. To get 150 × 30, triple 1,500, or 150 × 30 = 4,500. This is a first approximation. Improve this by realizing that 150 × 32 = 150 × 30 + 150 × 2. Since 150 × 2 = 300, you have 150 × 32 = 4,500 + 300 = 4,800.

4. 120 × 24: The first approximation is 100 × 24 = 2,400. Now compute 20 × 24 and add it to the 2,400. You know that 10 × 24 is 240, and therefore, 20 × 24 = 480. The answer is 120 × 24 = 2,400 + 480 = 2,880.

5. 340 × 71: First, 3 × 7 = 21. From this, you realize that 300 x 70 = 21,000. This is a first approximation. The first refinement is that 40 × 70 = 2,800. This you add to 21,000 to get 23,800. Finally, 340 × 71 = 340 × 70 + 340 × 1. You compute 340 × 1 = 340. Now, 340 × 71 = 23,800 + 340 = 24,140.

DIVISION

Mental division is much the same as mental multiplication. Simply separate numbers and approximate until you are close. Take $^{141}/_{17}$ as an example. Look for a number that, when multiplied by 17, gives 141 or close to it. You're looking for a number, N, such that $17 \times N = 141$. First multiply 17 by 10 to get 170. But 170 is more than 141; so you know that N is smaller than 10. Next divide 170 by 2 to get 85. But 85 is less than 141; so you know that N is greater than 5 and that N is somewhere between 5 and 10. Now try subtracting 17 from 170. This has the effect of multiplying 17 by 9 instead of by 10. So 170 minus 17 is 153 (170 − 20 = 150; plus 3 = 153) Now you're getting close, since 153 is close to 141. How close? Just 12 off. Therefore, you know that N is between 8 and 9. If you subtract another 17 from 153 you have, in effect, multiplied 17 × 8. This yields 136, and 136 is just 5 less than 141. Let's guess that N is 8.3. How close is that? The actual answer to 141/17 is 8.2941176, according to my calculator. Our guess was 8.3. Very close, indeed!

Here's the problem in steps.

(the problem)	141/17
(restating the problem)	$17 \times$ (what number?) $= 141$
(testing for 10)	$17 \times 10 = 170$; too big
(testing for 9)	$170 - 17 = 153$; too big
(testing for 8)	$153 - 17 = 136$; too small
(estimating)	$141/17 = 8.3$

Rule

To mentally divide large numbers (top number/bottom number), restate the problem as one of finding a number N such that N multiplied by the bottom number equals the top number. Guess at N, and then test N by multiplying by the bottom number. Continue testing and adjusting N until you reach the desired accuracy.

PROBLEMS

Do each division mentally to find an approximate answer.

1. 41/3

2. 291/17

3. 844/22

4. 1,466/190

ANSWERS

1. $41/3 = ?$ becomes $3 \times ? = 41$
(testing for 12) $3 \times 12 = 36$ (too small)
(testing for 15) $3 \times 15 = 45$ (too large)
(testing for 14) $3 \times 14 = 42$
(estimating) $41/3 = 13.8$ (actual answer $= 13.67$)

2. $291/17 = ?$ becomes $17 \times ? = 291$
(testing for 10) $17 \times 10 = 170$ (too small)
(testing for 20) $2 \times 170 = 340$ (too large)
(testing for 18) $17 \times 18 = 17 \times 20 - 17 \times 2 = 340 - 34 = 306$ (too large)
(testing for 17) $17 \times 17 = 306 - 17 = 289$ (too small)
(estimating) $291/17 = 17.1$ (actual answer $= 17.12$)

3. $844/22 = ?$ becomes $22 \times ? = 844$
(testing for 10) $22 \times 10 = 220$ (too small)
(testing for 40) $22 \times 40 = 4 \times (220) = 880$ (too large)
(testing for 39) $22 \times 39 = 880 - 22 = 858$ (too large)
(testing for 38) $22 \times 38 = 858 - 22 = 836$ (too small)
(estimating) $844/22 = 38.4$ (actual answer $= 38.36$)

4. 1,466/190 = ? becomes 190 × ? = 1,466
   (testing 10) 190 × 10 = 1,900 (too large)
   (testing 9) 190 × 9 = 1,900 − 190 = 1,710 (too large)
   (testing 8) 190 × 8 = 1,710 − 190 = 1,520 (too large)
   (estimating) 1,466/190 = 7.5 (actual answer = 7.72)

FIGURING OUT TIPS AT RESTAURANTS

Figuring tips at restaurants is such a common mental exercise that his entire section is devoted to it. Many people dread computing ips, believing it requires a complex mental calculation. Not so. You'll ·ncounter a little shortcut that simplifies the task.

First, let's agree that no one cares if the tip is calculated exactly to he penny. You don't care, the server doesn't care, and your luncheon riends don't care. All you really want is a good approximation. Sec-ond, let's agree that you need to think about only three tipping rates: 10%, 15%, and 20%. One of these three rates should apply to any of ·our tipping situations. If you are conservative, you may possibly tip ·nly 10%. If the service was good, you'll probably want to tip either 15% or 20%.

The first step is to divide the entire bill by 10. This is equivalent o calculating 10% of the bill. To divide the bill by 10, just move the ·ecimal point one place to the left. If the bill is $32.17, then 10%, ·r one tenth, is $3.21. Don't worry about the 7 on the end. If you're ipping only 10%, you might consider leaving $3.25. The 20% tip is ·lso easy to compute. Once you have a 10% tip computed, just double ·. Return to the $32.17 bill, and the $3.21 for 10%. Now double it to ·6.42. Again, don't worry about exact change; leave $6.25 or $6.50.

The 15% tip is not difficult at all. Once you have the 10% tip, in-·rease it by ½. Again, begin with $32.17, and get $3.21. Now mentally ·ivide $3.21 by 2 to get $1.60. Add this to the $3.21 to yield $4.81. ·nce again, you don't want to worry about pennies. If you start with ·32.17, you can approximate the 10% at $3.20 (ignoring the 1 cent); ·nd that ½ of $3.20 is $1.60, and add to get $4.80. Leaving $4.80 or ·5.00 is fine.

Computing tips in this manner avoids scribbling down figures on · napkin with a borrowed pencil and carrying out a long multiplica-·on problem. Choose your favorite percentage (10%, 15%, or 20%), and practice computing it at meals. In a short time, your friends will begin to look to you for the right tip. Ready to dine? Try the sample problems.

PROBLEMS

Estimate the appropriate tip for the following bill at 10%, 15%, and 20%. Your answer may be somewhat different from the answers shown, which have been rounded to the nearest 5 cents.

1. $4.62   4. $2.32
2. $16.81   5. $110.67
3. $37.19   6. $372.41

ANSWERS

	10%	15%	20%
1.	$0.45	$0.70	$0.90
2.	$1.70	$2.50	$3.35
3.	$3.70	$5.60	$7.40
4.	$0.25	$0.35	$0.50
5.	$11.10	$16.60	$22.10
6.	$37.25	$55.90	$74.50

WHAT HAVE YOU LEARNED?

1. Practice adding the digits 1 through 9 to any number that you can conveniently remember. Learn to add larger numbers that end in 0.

2. Rule: When mentally adding two large numbers, separate and add individual digits, beginning at the left, moving to the right, and ending with the ones digits.

3. Rule: To mentally subtract with large numbers, identify the smaller number, begin with the digits at the left, and subtract these digits

from the larger number. When finished, determine whether the answer should be positive or negative.

4. Usually, you can approximate the answers to problems instead of always trying to find the exact answer.

5. Rule: To mentally multiply two large numbers, study the numbers and then round or break them into parts to carry out easy multiplication. For more accuracy, adjust the answer by adding or subtracting portions missed due to rounding.

6. Rule: To mentally divide large numbers (top number/bottom number), restate the problem as one of finding a number N such that N multiplied by the bottom number equals the top number. Guess at N, and then test N by multiplying by the bottom number. Continue testing and adjusting N until you reach the desired accuracy.

You have also learned a simple procedure to compute restaurant tips at 10%, 15%, and 20%.

CONCLUSION

Mental arithmetic requires practice. The better you become at it, the more fun it becomes, and the more useful it is. One nice feature of mental arithmetic is that no one can tell you're doing it—it's completely private. While in conversation with friends, little problems sometimes pop up and are swiftly ignored. You can mentally find the answer without anyone knowing what you're doing and then spring the answer on them. They'll look at you with eyes wide in wonderment. They'll ask, "How did you do that?"

Try to practice mental arithmetic for both curiosity and necessity. Take some problem from the environment: How many cubic feet of wood are there in that tree out back? If it snows to a depth of one foot, how many pounds of snow will be on your roof? How many gallons of water will fill little Cindy's portable swimming pool? Use approximation and guesses. Once you begin, you'll have a pleasurable pastime that can exercise your mind and drive away boredom.

Being proficient in mental arithmetic is not an absolute requirement for learning more advanced mathematics. Even if you have difficulty with it, don't hesitate to study and practice more math. And now, on to the big M—MONEY!

CHAPTER 20

MONEY POWER

Now that you have mastered the basic techniques of arithmetic, and the thought of math no longer makes your blood freeze, you are ready to apply this knowledge to controlling your finances. You have already read of matters that affect the pocketbook. In Chapter 16 you explored electrical usage, gas consumption, and unit pricing. Chapter 18 covered interest rates. In this chapter you'll review how to balance your checkbook, manage your receipts and bills, analyze your food budget, project your budget, and avoid deficit spending. This chapter concludes with a few remarks about investments.

BALANCING YOUR CHECKBOOK

Balancing your checkbook is an important step in controlling your personal budget. If you aren't sure how much money you have, it's difficult to plan how to spend it. When writing checks, remember to enter into your check ledger the check number, what the check was for (very important when tax time comes around), and the amount of the check. Always subtract the current check amount from the balance to get a new balance (Figure 13).

Most people have little difficulty listing their checks. It's when they match their balances against the bank statements that they run into trouble. The basic task in balancing your checkbook is to adjust both the bank's ending balance and your ending balance and then compare them to see that they agree. Here are the four basic steps in this process.

Check No.	Date	ITEM	Returned from bank	Balance
				1,271.01
721	4/9/90	Car Payment	✓	146.98
			Bal	1,124.03
722	4/9/90	Life Insurance - Year	✓	276.50
			Bal	847.53
	4/12/90	Deposit of Monthly Paycheck		1,417.62
			Bal	2,265.15
723	4/21/90	May Mortgage Payment		762.14
			Bal	1,503.01
724	4/24/90	Lumber for New Deck		491.37
			Bal	1,011.64
725	4/25/90	Groceries		262.19
			Bal	749.45
			Bal	
			Bal	
			Bal	

Figure 13.   Entering checks in a checkbook

1. When you receive your bank statement, go through your checks and match the check numbers against the numbers listed in the statement. This ensures that one or more of the checks the bank has listed has not been misplaced by it or by you. Also review all checks to make sure the correct amounts have been read from them.

2. Adjust the bank's balance on the bank statement:

   A. Page through your check ledger, and find any deposits you have made that the bank has not shown. Add these deposits to the bank's ending balance (Figure 14).

   B. Page through your check ledger, and put an *X* or other mark beside each check that the bank has processed. When you have finished, write down the amounts of all those checks you have written that the bank has not yet cleared. Add the amounts to get a dollar figure, and subtract this amount from the bank's balance on the statement. Now the bank's balance should be correct.

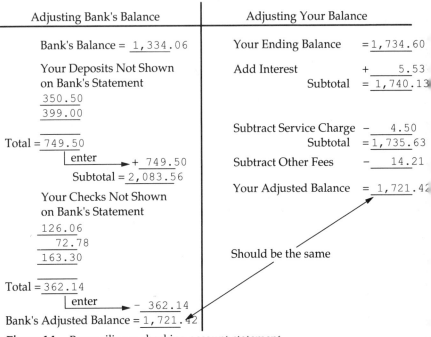

Adjusting Bank's Balance	Adjusting Your Balance

Bank's Balance = 1,334.06

Your Deposits Not Shown
on Bank's Statement
350.50
399.00

Total = 749.50

⌐enter ────► + 749.50

Subtotal = 2,083.56

Your Checks Not Shown
on Bank's Statement
126.06
72.78
163.30

Total = 362.14

⌐enter ────► – 362.14

Bank's Adjusted Balance = 1,721.42

Your Ending Balance = 1,734.60

Add Interest   + 5.53
Subtotal   = 1,740.13

Subtract Service Charge – 4.50
Subtotal   = 1,735.63

Subtract Other Fees   – 14.21

Your Adjusted Balance = 1,721.42

Should be the same

Figure 14.   Reconciling a checking account statement

3. Adjust your checkbook balance:

A. If you have an interest-bearing checking account, you may have earned interest. Check the bank's statement. If you have been credited with interest, add this amount to your balance.

B. Subtract from your balance the bank's service charge (if any) and any other charges the bank has made against your account. These will all be listed on the bank's statement.

4. Compare the bank's adjusted balance to your adjusted balance. If they are the same, you are finished.

What if your adjusted balance is different from the bank's? There could be several sources of error.

1. While writing a check, you may have improperly subtracted the amount of the check from the old balance to get the new balance. This is a common error. From that point in your ledger at which

you balanced your account the previous month, recheck each entry to make sure you performed the subtraction correctly.

2. You may have incorrectly added deposits into your check ledger. Recheck each deposit entry.

3. You may have incorrectly added the deposits not shown by the bank into the bank's balance. Recheck this addition.

4. You may have added outstanding checks incorrectly. Add them again.

5. You may have improperly subtracted the service fee and other fees from your balance. Recheck this subtraction.

6. If these steps fail to find the error, you might have improperly balanced your ledger against the previous month's bank statement. If you succeeded in balancing your account last month, this probably is not the error. However, eliminating the possibility requires going back and rebalancing the previous month's statement—a messy proposition.

If this checklist does not pinpoint the error, the bank may have made a mistake. This is unlikely, but it can happen. You can double-check the bank's arithmetic on the statement, adding and subtracting deposits and checks processed against their beginning and ending balances. If you can't find the error, you still have two choices. If the difference is small, you can assume you made a mistake and adjust your checkbook balance to agree with the bank's. Or you can take your checkbook and statement to the bank for a review. Of course, if the disagreement is large and you can't find its source, you should *always* take your statement in for review.

Pay attention to service fees: Always examine any charges against your account. The ideal situation is to have sufficient money in the bank, to pay no service fees, and to actually earn interest against your balance. However, many of us simply can't afford to keep that much in our accounts. If the bank's service charges are running $5.00 or less, and you write a dozen or more checks per month, you are paying $60.00 per year for the convenience of writing checks. This is not excessive. However, if you are paying substantially more, you should consider either changing the kind of checking account you have or finding another bank. Try some problems.

PROBLEMS

1. Examine the partial check ledger. Complete the entries by computing the missing balances until you have a current balance.

Check No.	Item		Amount
		Balance	$1,756.72
2011	Car Insurance		$742.40
		Balance	$1,014.32
2012	Grocery Store		78.14
		Balance	$963.18
2013	Antique Auction		71.94
		Balance	
2015	Electrical Bill		114.65
		Balance	
2016	John's Birthday		24.91
		Balance	
2017	Dentist		42.15
		Balance	

2. For the account represented in Problem 1, the bank statement says the service fee for the month was $4.90 and that the current balance is $677.63. After adjusting for the service charge, does this agree with the balance you computed in Problem 1? If not, was there an error in the listing in Problem 1?

ANSWERS

1. By filling in the missing balances, the current balance will be $709.53.

2. After subtracting the service fee of $4.90 from the last balance in Problem 1, the new balance is $704.63. This is far from the bank's balance of $677.63. By checking the subtraction in Problem 1, you discover that $1,014.32 minus $78.14 is not $963.18 but $936.18.

CHASING OUT THE MONEY CHANGERS

The next issue of money handling is to regain control of your bills. Begin by checking your receipts from purchases, especially any receipt that is handwritten. You will discover that many handwritten restaurant bills are in error. The error will almost always be against you and in favor of the establishment. Servers sometimes accidentally overcharge. This typically occurs in one of three ways (or any combination of them): They charge you for the wrong items, put down an item more than once, or add incorrectly. When reviewing your bill, first check each item listed; then check the addition. The server won't care; he or she is too busy waiting on other tables to worry about your checking the bill.

"Why," says the critic, "should I waste my time checking bills and receipts for a few pennies, when I could be using that time to make more money—real money?" The answer is simple: It is easier to save a dollar than to earn one. Look at a simple example of how saving money compares to earning money. At the grocery store, tomatoes are marked three for a dollar, and you put three in your cart. The clerk accidentally rings them up at two for a dollar and charges you $1.50. You stop him and point out his error. He adjusts the total by 50 cents. You've just saved yourself 50 cents in the space of twenty seconds. That's $1.50 per minute, or $90 per hour. When was the last time you earned $90 per hour? In terms of an hourly rate, you can't beat checking for mistakes made against you by others.

There is another reason to look for and catch these kinds of mistakes besides the money you save. When you catch a mistake and save yourself money, you feel good about yourself; you feel in control again. You have the sense that it is you, not someone else, who directs and manages your affairs.

Check even machine-generated receipts, not necessarily for the addition but for the items included and the cost per item. Carry out your review in an open fashion. Clerks and salespeople are used to it. After all, it's your money.

Besides checking your daily purchases, you should also review your bills. These can include your mortgage or rent, heating, electricity, water, garbage carting, sewer, insurance, food, transportation, and clothing. Any of these bills can yield potential savings. Over time you can save hundreds or even thousands of dollars. Your first step is simple. When you get a bill, open it and read it. From what's written, try to determine how the amount you currently owe was calculated. The statement frequently includes a unit or rate price multiplied by the amount of your purchase. When you understand how the bill was generated, ask yourself if there is any way to control or reduce that bill without affecting your quality of life. If you don't understand the bill, or if you find it confusing, call the billing company for an explanation. After a few minutes of discussion with a company representative, you may discover that you are paying at a premium rate, and that, with a slight change in billing, you can save money.

YOUR FOOD BUDGET

Some people are naturally meticulous about food purchases, buying only necessities and using coupons. If you are already one of these people, your food bill is probably under control. However, if you're not paying attention to food purchases, you can probably save money. For a family of five, you could easily be spending $500 to $600 per month. Since most of us shop at supermarkets, this dollar figure may also include certain nonfood items such as cleaning compounds, light bulbs, minor hardware items, and so on. If you can save 10% on this food bill, you save between $50 and $60 per month, or $600 to $720 per year. If you have a larger family or live in a metropolitan area with an especially high cost of living, your food bill can be even more and the savings even greater.

The first step is to perform a food-purchasing analysis. Many modern stores use cash registers that print on your receipts the name of the item purchased, the units or amounts you purchased, and the price you paid. This is a boon to those of us who are interested in tracking our buying habits. If your favorite grocery store doesn't use such a cash register, try to find a store you like that does. If this is not possible, you're going to have to keep a handwritten record of everything you buy. It's a dirty job, but someone has to do it.

Retain your grocery store cash-register receipts and keep a running list of purchases in various categories for two or four months. Add the amounts in each category to determine how you are spending your money. Table 20.1 is an example of a representative food budget for a family of five, including two adults, one teenager, and two small children.

Remember that individual food budgets vary greatly, and Table 20.1 is given to demonstrate the budgeting process rather than what your budget should or would look like. Note that, in Table 20.1, nonfood items represent $159.83, or 26.6% of the total. (Pet food has been classified as nonfood since it is a nonfood for humans). Meat was the single largest item among the food categories, accounting for 19% of

Table 20.1 Purchasing Analysis for Food

	Cost Per Month	Cost Percent
Food		
Meat (beef, bacon, chicken)	$113.96	19.0%
Drinks (soda, coffee, juice, wine)	86.40	14.4%
Fruit and vegetables	62.27	10.4%
Dairy (milk, butter, cheese, cream, eggs, yogurt, cottage cheese)	53.31	8.9%
Snacks and dessert (cookies, ice cream, chips, pudding, crackers, gum, popcorn, candy)	42.47	7.1%
Bakery (bread, stuffing, buns)	29.83	5.0%
Toppings and sauces (salad dressing, honey, catsup, maple syrup)	23.73	4.0%
Pasta, soups, cereals	17.53	2.9%
Misc. cooking supplies (cooking oil, salt, flour, sugar)	10.67	1.8%
Nonfood		
Pet food (2 dogs and 2 cats)	70.26	11.7%
Detergent (detergent, soap, sponges)	19.41	3.2%
Personal hygiene (hand soap, shampoo)	17.61	2.9%
Paper products (napkins, bags, towels)	14.27	2.4%
Clothing	6.39	1.1%
Miscellaneous hardware (batteries, tape)	5.66	0.9%
Miscellaneous (film, medicine, flowers, toys)	26.23	4.4%
Total	$600.00	

the entire budget. The second most expensive category was Drinks followed by Pet food. These three categories accounted for 45% of the entire expenditure. Therefore, in trying to decrease the food expenditure (without decreasing the quality or "fun value" of the food), you should first consider ways to save on meat, drinks, and pet food. You are looking at the largest expenditures first, because any improvement you make in these will have the biggest impact on your savings.

You can save on numerous items by buying in bulk and shopping at discount outlets. Some items—such as milk, fresh vegetables, and eggs—cannot be stored for long periods. However, many other items can be purchased in bulk, stored in your home, and used as needed. Most metropolitan areas offer one or more large warehouse outlets for buying in bulk. Some of these require an initial membership fee, but if you use the membership, the savings from the first few shopping trips will more than make up for this extra cost. However, just because a warehouse outlet carries an item, you can't assume it is cheaper than your favorite grocery store. You must compare prices.

Meat can be purchased in bulk and frozen. If you do not own a freezer and can't afford to buy a new one, consider buying a used one. They are relatively inexpensive and last for years. Another option is to purchase one with a neighbor or relative and share both the space and the expense. As you can see in Table 20.1, a 10% savings on meat translates to a yearly savings of $137. A 20% savings in this category yields a savings of $274 per year.

Drinks (coffee, soft drinks, frozen juices) can also be purchased from a warehouse outlet and stored. Dog and cat food can be stored easily; canned pet foods and packaged pet foods need no refrigeration. If you can save 10% on the three largest items, you save $325 per year—enough to purchase a freezer. If you can stretch your savings to 20% on these items, you realize $650 per year. Other items should not be ignored. In fact, any item that is available in bulk and can be stored should be considered.

Sometimes you have to be careful with your money-saving alternatives. Since meat occupied 19% in the budget, you might be tempted to take up hunting to curtail your meat cost. However, meat that is acquired through hunting is generally more expensive than if it were purchased—just think of the licenses, ammunition, guns, and so on. Likewise, if you are tempted to start a garden to offset your fruit and vegetable costs, you must take into account the cost of equipment

ertilizer, and seeds. Also, remember that fruit and vegetables represented only 10% of the food budget in Table 20.1. Is it worth all that
work, plus the expense, to save on fruit and vegetables? If you love
gardening, then it may make sense.

─────────

FORECASTING YOUR BUDGET

One substantial benefit of proficiency in arithmetic is being able to
develop a meaningful household budget. Take a look at a hypothetical
budget for a young married couple, Stan and Lisa, who have two
children, John, 5, and Mary, $1\frac{1}{2}$. The budget will list income and
expenditures for a typical month for the couple and then forecast
their budget for one year. Since this budget process generally requires
a number of revisions, it is helpful to prepare a form and then make
a number of copies of the form.

The budget in Table 20.2 may not match your personal budget, but
remember that it is presented to show the process.

Table 20.2 Income and Expenses for Stan and Lisa

Net income	
Stan's work as a backhoe driver	$1,650
Lisa's part-time work	320
Total income	$1,970
Expenses	
Mortgage	$ 520
Telephone/newspaper	49
Electricity	70
Food	450
Car insurance (10 months)	65
Gasoline	90
Home maintenance fund	25
Stan's allowance	100
Lisa's allowance	100
Credit card #1 (average)	50
Credit card #2 (average)	50
John's summer school (2 months)	75
Summer vacation (August only)	1,000
Life insurance (April and October)	238
Medical insurance (6 months)	112
Car registration (June)	110
Christmas (December)	700

As you can see, some of the expenses are monthly, whereas others are due only in certain months. In addition, Stan and Lisa wan
to start a small long-term savings program that will protect them ir
emergencies and allow them to make a modest investment.

Table 20.3 shows how the budget table is to be set up for Jan
uary and the beginning of February. (But you can actually begin you:
budget during any month of the year.)

Stan and Lisa's budget table, when finished, will contain 25 line
and 14 columns. The first column is a list of the various items, the
next 12 are for each of the 12 months of the coming year, and the
last column is for the yearly total.

You should notice that we've added several additional lines in Ta
ble 20.3. First, we have added a line entitled *Cash reserve* (Line 1) and ;

Table 20.3 Beginning Budget Plan for Stan and Lisa

		January	February
1	Cash reserve	$ 0	$ 239
	Income:		
2	Stan's work	1,650	1,650
3	Lisa's work	320	320
4	Total income available	1,970	2,209
	Expenses:		
5	Mortgage	520	
6	Telephone/newspaper	49	
7	Electricity	70	
8	Food	450	
9	Car insurance	65	
10	Gasoline	90	
11	Home maintenance fund	25	
12	Stan's allowance	100	
13	Lisa's allowance	100	
14	Credit card #1	50	
15	Credit card #2	50	
16	John's summer school	0	
17	Summer vacation	0	
18	Life insurance	0	
19	Dental/medical	112	
20	Car registration	0	
21	Christmas	0	
22	To savings account	50	
23	Total expenses	1,731	
24	To carry forward as cash reserve	239	
25	(Total Savings Account)	50	

line entitled *To carry forward as cash reserve* (Line 24). The principle of carrying cash or resources forward is what really makes the budget-planning process work. During some months, Stan and Lisa's combined income is greater than their expenses. However, during other months, their expenditures exceed their income. By carrying cash forward from flush months to cover lean months, they will avoid serious trouble.

In Table 20.3, in the column under January, you subtract the total expenditures ($1,731) from the total income ($1,970) to get $239. Stan and Lisa will carry this $239 forward as a cash reserve to help cover months when their income is deficient. Therefore, you write the $239 into Row 1 under the month of February. In February and the following months, you add Rows 1, 2, and 3 to get Total Income Available (Line 4). Each time you get a surplus, you copy that surplus amount from Line 24 to Line 1 for the next month.

Table 20.4 shows the entire year's budget. Notice that in the summer months when Lisa isn't working at her part-time job, their total income decreases. Yet, it is during the summer that they incur additional expenses. If they didn't carry forward a cash reserve, they would have to borrow money each summer to go on vacation. This demonstrates an interesting fact about family budgets. People generally don't get into budget trouble when they're short of cash. During such times, they usually cut back spending and try to get by. It's when people have extra cash that they are likely to spend too freely and neglect to carry money forward for lean months.

To prepare your own budget, list all income and expenses as Lisa and Stan did. With pencil, draw sufficient columns and rows on a legal pad or wide computer paper. Once you have the months labeled and the items included in the left column, make copies of this sheet (it saves you from having to keep drawing all those columns!) because you might have to create an adjusted budget.

Fill in January, and carry your excess cash forward to February. Once you have completed February, go to March. If you reach a month when you don't have enough money, even with a cash reserve, you must go back to previous months and reduce expenses enough to increase your cash reserve (or you must increase income). Several attempts may be required to plan the kind of budget you want. If you have a computer and a spread-sheet program, the task is much easier,

Table 20.4 Twelve Month Budget Plan

		January	February	March	April	May
1	Cash reserve	0	239	590	894	1,072
	Income:					
2	Stan's work	1,650	1,650	1,650	1,650	1,650
3	Lisa's work	320	320	320	320	320
4	Total income available	1,970	2,209	2,560	2,864	3,042
	Expenses:					
5	Mortgage	520	520	520	520	520
6	Telephone/newspaper	49	49	49	49	49
7	Electricity	70	70	70	70	70
8	Food	450	450	450	450	450
9	Car insurance	65	65	0	0	65
10	Gasoline	90	90	90	90	90
11	Home maintenance fund	25	25	25	25	25
12	Stan's Allowance	100	100	100	100	100
13	Lisa's Allowance	100	100	100	100	100
14	Credit card #1	50	50	50	50	50
15	Credit card #2	50	50	50	50	50
16	John's summer school	0	0	0	0	0
17	Summer vacation	0	0	0	0	0
18	Life insurance	0	0	0	238	0
19	Dental/medical	112	0	112	0	112
20	Car registration	0	0	0	0	0
21	Christmas	0	0	0	0	0
22	To savings account	50	50	50	50	50
23	Total expenses	1,731	1,619	1,666	1,792	1,731
24	To carry forward as cash reserve	239	590	894	1,072	1,311
25	(Total Savings Account)	50	100	150	200	250

because the computer can make all the revisions when you adjust figures in the budget.

Remember that a budget is only a planning device. Your actual income and expenditures will occasionally vary from the figures in your budget because of unforeseen changes in your living situation. However, as you learn to use a budget plan, you will identify potential problems and learn to plan for special events such as vacations and for the acquisition of specialty items. One more tip: Don't let a whole year go by before preparing another complete budget. Instead, revise your budget every three to six months, extending the projection twelve months each time.

June	July	August	September	October	November	December	Total
1,311	1,157	1,001	32	31	144	383	
1,650	1,650	1,650	1,650	1,650	1,650	1,650	19,800
0	0	0	80	320	320	320	2,640
2,961	2,807	2,651	1,762	2,001	2,114	2,353	
520	520	520	520	520	520	520	6,240
49	49	49	49	49	49	49	588
70	70	70	70	70	70	70	840
450	450	450	450	450	450	450	5,400
65	65	65	65	65	65	65	650
90	90	90	90	90	90	90	1,080
25	25	25	25	25	25	25	300
100	100	100	100	100	100	100	1,200
100	100	100	100	100	100	100	1,200
50	50	50	50	50	50	50	600
50	50	50	50	50	50	50	600
75	75	0	0	0	0	0	150
0	0	1,000	0	0	0	0	1,000
0	0	0	0	238	0	0	476
0	112	0	112	0	112	0	672
110	0	0	0	0	0	0	110
0	0	0	0	0	0	700	700
50	50	50	50	50	50	50	600
1,804	1,806	2,619	1,731	1,857	1,731	2,319	
1,157	1,001	32	31	144	383	34	
300	350	400	450	500	550	600	600

DEFICIT SPENDING OR SAVINGS

We all know that it is sound money management to develop a long-term savings plan. However, it can be so hard! It's hard for me, and it's hard for almost everyone I know. How many times have I saved a small amount and then been faced with a completely unexpected but necessary expense? Saving money for a time far off in the future is psychologically difficult. You see money going into savings and feel deprived because you can't enjoy it right away.

Sometimes you may not only neglect to save, but actually spend *more* money and then, to make matters worse, charge on your credit cards. Each month becomes a struggle not to overspend, not to get deeper in debt.

Let's observe two couples, Delores and Dave Debtor and Susan and Steve Saver. You'll track their spending and saving habits over twenty years and see which couple actually gets to spend more on themselves. Each couple will earn $30,000 per year. Delores and Dave will overuse their credit cards by 5% of their earned income per year. Susan and Steve will save 5% of their earned income per year. Delores and Dave will have to pay 15% interest on their debt each year, which they will have to subtract from their spendable income, while Susan and Steve will earn 7.5% on their savings, which they will add to their spendable income.

After twenty years, what has happened to the Savers and the Debtors? Look at Table 20.5, which shows the accumulated debt, the accumulated savings, the yearly expenditures, and the accumulated spending of both couples.

The Debtors' income has eroded because of interest on their debt, while the Savers' usable income has increased from their interest

Table 20.5 Savings, Debt, and Spending of Two Couples

Year	Accum. Debt Debtors	Accum. Savings Savers	Yearly Spending Debtors	Yearly Spending Savers	Accum. Spending Debtors	Accum. Spending Savers
1	$ 1,500	$ 1,500	$31,275	$28,612	$ 31,275	$ 28,61
2	3,000	3,000	31,050	28,725	62,325	57,33
3	4,500	4,500	30,825	28,838	93,150	86,17
4	6,000	6,000	30,600	28,950	123,750	115,12
5	7,500	7,500	30,375	29,062	154,125	144,18
6	9,000	9,000	30,150	29,175	184,275	173,36
7	10,500	10,500	29,925	29,288	214,200	202,65
8	12,000	12,000	29,700	29,400	243,900	232,05
9	13,500	13,500	29,475	29,512	273,375	261,56
10	15,000	15,000	29,250	29,625	302,625	291,18
11	16,500	16,500	29,025	29,738	331,650	320,92
12	18,000	18,000	28,800	29,850	360,450	350,77
13	19,500	19,500	28,575	29,962	389,025	380,73
14	21,000	21,000	28,350	30,075	417,375	410,81
15	22,500	22,500	28,125	30,188	445,500	441,00
16	24,000	24,000	27,900	30,300	473,400	471,30
17	25,500	25,500	27,675	30,412	501,075	501,71
18	27,000	27,000	27,450	30,525	528,525	532,23
19	28,500	28,500	27,225	30,638	555,750	562,87
20	30,000	30,000	27,000	30,750	582,750	593,62

earned from savings. At first, the Debtors are enjoying more money to spend, but with time, that advantage is eaten away. By the seventh year, the Debtors' spendable income has dropped below their base salary of $30,000. Hence, they no longer derive any advantage from overspending their income by 5% because of interest. By the ninth year, the Savers' spendable income has exceeded that of the Debtors. At the end of twenty years, Susan and Steve Saver have actually spent almost $11,000 more than Dave and Delores Debtor. In addition, the Savers have a $30,000 nest egg, while their friends, the Debtors, owe $30,000.

Why have we gone through this example? We have demonstrated the simple principle that people who save can actually spend more over the long run than people who don't save. Some people say, "I can't afford to save. I'm barely making it now." The truth is, you can't afford not to save. Now you have the tools to carry out a budget-and-savings plan.

NOTES CONCERNING INVESTMENTS

Now that you are saving money, how do you ensure that your money is earning you more money and is safe? First, let the accumulated interest return to the savings fund. This interest will be so low at first, that it will make no discernible difference in your standard of living. Turn the interest back into the savings fund (not necessarily a savings account), and your interest will begin to earn interest of its own. Check Table 20.6 to see what a savings fund can grow to at different interest rates.

Table 20.6 Examples of Various Savings Programs

	20 Years	30 Years	40 Years
$10 per month at 7.5% interest	$ 5,391	$ 12,873	$ 28,293
$10 per month at 10% interest	7,217	20,726	55,767
$10 per month at 12.5% interest	9,736	33,908	112,403
$25 per month at 7.5% interest	13,479	32,183	70,734
$25 per month at 10% interest	18,042	51,816	139,417
$25 per month at 12.5% interest	24,342	84,770	281,007
$50 per month at 7.5% interest	26,957	64,366	141,467
$50 per month at 10% interest	36,083	103,631	278,833
$50 per month at 12.5% interest	48,680	169,541	562,015

From Table 20.6, you can see that a modest $10-per-month invest-
ment that grows at only 7.5% a year will yield more than $5,000 in
twenty years. At the other extreme, $50 a month at 12.5% interest
will grow to more than $500,000 in forty years. This means that a
25-year-old can put away $500,000 by the time he or she retires. It
also demonstrates that young people have in abundance a valuable
commodity that older people have in short supply—time. Start to save
when you are young: beginning is more important than the amount
you actually save.

The interest rate your savings fund earns is also important. How do
you ensure that your savings earn a reasonable rate while remaining
safe? There are many books available on sound investment programs.
This book teaches you only a few general guidelines. First, the less
you know about investments in general, the more conservative your
investment should be. Never enter into an investment scheme that
you don't understand or that requires you to place all your trust in
someone else. Get-rich-quick schemes that are just "too good to be
true" are just that—too good to be true. Stay away from them.

Here's a quick review of the various options you have for saving
money and earning money from savings.

1. *The bank.* Your bank offers several good savings devices that include interest-bearing savings accounts, certificates of deposit, U.S
   Savings Bonds, money-market savings, and time deposits. These
   savings devices will generally earn you from 4% to 7% in yearly
   interest. This isn't much. However, these investments are relatively
   safe and, depending on the one you choose, make it easy to have
   access to your money. For small amounts of money that you may
   need quickly, they offer a good investment selection. The personnel
   at your bank will be pleased to review each of these options with
   you.

2. *The stock market.* If you want to earn a higher return on your investment, the stock market may be the place for you. But stand
   warned, it is also more risky. This points out the general rule that
   the investment that offers more earnings will also expose you to
   more risk. If the stock market interests you, read several books
   about it and then select a broker. You should plan on a return of
   10% to 15% a year to make the extra risk worthwhile.

3. *Real estate.* Of course you've heard of the people who used to live
   down the street, dabbled in real estate, and now live on the Riviera

It happens, but it happens rarely. The best real-estate investment you can make is to purchase your own home. While your investment grows, you live in it, write it off your taxes, and personally protect it from damage. By all means, if you do not now own your home, read some good books on home buying and then buy one. What about other real estate? If you are interested in entering the business, do your homework first.

4. *That special scheme.* This includes limited partnerships, lending money to your brother-in-law's nephew, investing in art through a friend, supporting a gold-prospecting expedition into Uganda, and any of the other harebrained ideas people come up with to take your money. Don't do it. Don't get into any special deals you hear from others. Don't lend money to friends or family. If they truly need it—give it as a gift and write it off. If they are honest and loyal, they will reciprocate. If not, you haven't lost anything from your investment fund.

Of the four areas outlined, the first three are legitimate and offer sound investment opportunities. You should proceed in the following manner: Always have a few hundred dollars of savings available in cash in case of an emergency (but not for an impulse spending spree). With this buffer you can avoid borrowing small amounts at high interest rates. Next make a second investment of several thousand dollars in one of the options offered by the bank. This will be money that is harder to get, relatively safe, but available in a few days in case of a catastrophe.

If there is money left, decide if you want to go for stocks (including mutual funds) or real estate. If neither, you'll have to be satisfied with the safe harbor offered by the bank. If you want a greater return, start by buying good stocks with a history of growth. If you are tempted to take big risks with your money, limit the high-risk investment to no more that 10% of your investment funds. Do not put all your money into one investment. But do not keep moving it around needlessly, since this costs you in brokerage fees and real-estate costs.

Most important, start saving now!

WHAT HAVE YOU LEARNED?

1. It is easier to save a dollar than to earn one.
2. Keep your checkbook balanced.

3. Review your receipts to catch any errors. Review each of your household and living-expense bills, and ask yourself if there are ways you can save. Pay special attention to the large items such as the mortgage, utilities, and food. When possible, buy in bulk.

4. Forecast your personal budget for a year in advance. Update this forecast every three to six months. Include a savings plan within your budget. Remember that people who save have more to spend on themselves than people who incur debt.

5. When you have a savings plan, keep several hundred dollars in cash for small emergencies. Have several thousand dollars in a bank savings program where you can get at it with only a few days notice in case of a calamity. Find another investment program for long-term growth. Limit high-risk investments to 10% or less of your investment funds. Never invest in get-rich-quick schemes.

Next up—those vexing charts and graphs.

CHAPTER 21

CHARTS AND GRAPHS

The purpose of a graph is to present a visual illustration of numerical data. A picture is often easier to grasp and remember than the data, and with a graph, the information contained in the data can be quickly absorbed and retained.

THE BAR GRAPH

Graphs are popular communication devices in both business and government and are used constantly by authors to emphasize or support their ideas. The basic graph is a picture of varying magnitudes that the author wants the reader to compare. The simplest graph is the bar graph. Graphs have three separate elements.

A scale

Categories

Associated values or data

Figure 15 is a vertical bar graph because the scale is vertical. The various categories are along the bottom. The associated table of values is given in Table 21.1.

The vertical scale at the left of Figure 15 has been marked from 0 to 40 gallons. Since 26.3 gallons of milk were consumed per capita in 1986, the first bar, over milk, has been drawn to a height corresponding to 26.3 on the scale.

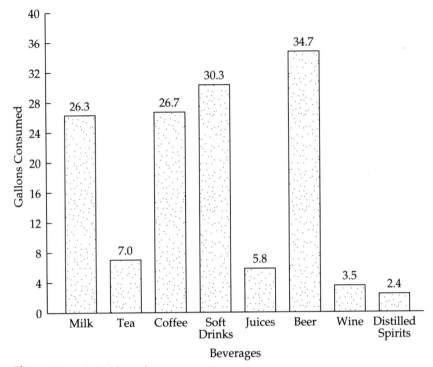

Figure 15. Vertical bar graph: per capita consumption of beverages, by type for 1986. *Source:* Bureau of the Census, *Statistical Abstract Of The United States, 1990* (Washington D.C.: GPO, 1990) 125.

Table 21.1 U.S. Per Capita Consumption of Beverages, by Type, 1986

Type of Beverage	Average Gallons Consumed Per Year Per Capita
Milk	26.3
Tea	7.0
Coffee	26.7
Soft drinks	30.3
Juices	5.8
Beer	34.7
Wine	3.5
Distilled spirits	2.4

Source: Bureau of the Census, *Statistical Abstract of the United States—1990,* (Washington, D.C., GPO, 1990) 125.

What does the graph in Figure 15 tell you? Well, it shows clearly that Americans drank more soft drinks than coffee or milk and more beer than any other beverage. You can also see from Figure 15 the comparable amounts of other beverages consumed on a per capita basis. Figure 15 is a well-constructed graph; the message is understandable at a single glance.

The bar graph in Figure 16 is from the 1989 Population Profile of the United States by the Bureau of the Census. It compares the percent of the United States population who are college graduates for both sexes and for different age groups. Notice that the scale has been omitted. This focuses the eye more quickly on the bars. Table 21.2 contains an associated table of values.

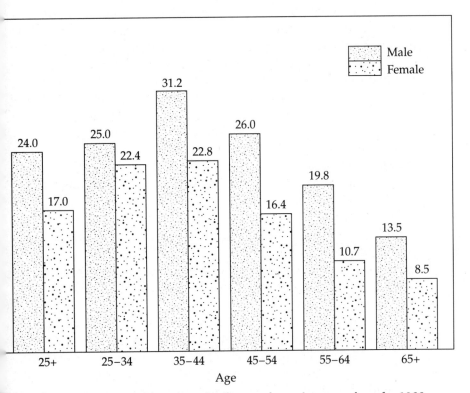

Figure 16. Vertical bar graph: percent of college graduates by age and sex for 1988.
Source: Bureau of the Census, *Population Profile of the United States* (Washington D.C.: GPO, 1989) 22.

Table 21.2 Percentages of College
Graduates, by Age and Sex for 1988

Age Group	Men	Women
25+	24.0	17.0
25–34	25.0	22.4
35–44	31.2	22.8
45–54	26.0	16.4
55–64	19.8	10.7
65+	13.5	8.5

Source: Bureau of the Census, *Population Profile of the United States, 1989,* (Washington, D.C., GPO, 1989) 22.

Figure 16 is a little confusing because it is not immediately evident from the graph what the various percentage figures represent. Do they represent, for example, the percent of all college males who are within a particular age group, or do they represent the percent of the total male population within that age group who are college graduates? It turns out to be the second interpretation. Hence, the 24.0 over the first bar on the left means that of all males in the entire population of the United States 25 years and older, 24% are college graduates. For females, 17.0% are graduates. Without clear labeling, this graph is a poor communicator.

Figure 16 tells you that a greater percentage of males than females in the population are college graduates in all age groups and that a higher percentage of younger people (ages 25-54) have degrees than do older people (ages 55+).

Some bar graphs are horizontal; the scale is at the bottom of the graph, and the various categories are at the left. Figure 17 is an example of a horizontal bar graph. It shows the average weekly food cost for different sizes of families for 1988. The scale is at the bottom, and the categories are at the left.

The graph in Figure 17 shows that larger families pay more for food than smaller families. Of course, this should be no surprise. The graph also shows that couples more than 51 years of age spend only slightly less than younger couples. From the graph, you can see that a family of four with two small children pays an average of $94.90 per week, or approximately $410 per month for food.

The bar graph can be used in a number of home-budgeting projects. You can keep track of various expenses by week or by month for an

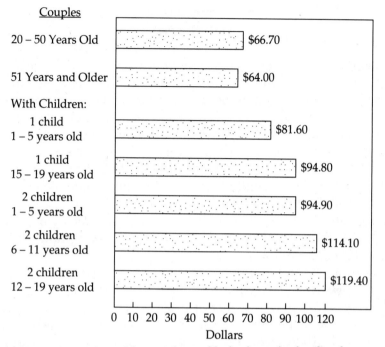

Couples

20 – 50 Years Old — $66.70

51 Years and Older — $64.00

With Children:
1 child
1 – 5 years old — $81.60

1 child
15 – 19 years old — $94.80

2 children
1 – 5 years old — $94.90

2 children
6 – 11 years old — $114.10

2 children
12 – 19 years old — $119.40

Dollars (0 10 20 30 40 50 60 70 80 90 100 120)

Figure 17. Horizontal bar graph: weekly food cost for families, by type of family for 1988. *Source:* Bureau of the Census, *Statistical Abstract Of The United States, 1990* (Washington D.C.: GPO, 1990) 484.

entire year. At the end of the year, you can draw a bar graph to see how each expense changes by season. Do your food costs vary by month? Do your heating and electrical bills change during the year? What about water usage? Understanding how your spending for such necessities varies during the year can help you to plan a more accurate budget.

To make a graph, begin with a table of values, such as Table 21.1. Once you have the table containing the data and the categories, determine the scale. Always begin with 0 as the lower limit and then choose an upper limit greater than the greatest number in the table. The greatest number in Table 21.1 is 34.7, so the scale goes from 0 to 40. Now draw bars of the correct magnitudes.

To make a horizontal bar graph, place the scale at the bottom and the categories at the left, as in Figure 17.

PIE GRAPHS

The pie graph is based on a circle, in which the area of the circle represents 100% of something. The various categories are represented as slices of pie. Like bar graphs, pie graphs have categories and associated tables of values, but the scale is replaced by the circle. Figure 18 is a pie graph that shows the percentages of homeowners with varying kinds of debt for 1986. Everyone in this graph is a homeowner—that is the 100%. This graph tells you that 26% of United States homeowners had neither mortgage nor consumer installment debt. Almost half the homeowners (44.4%) had both kinds of debt.

Pie graphs have a rather limited use and are not as popular as bar graphs. However, they are convenient for showing how a part of something relates to the whole.

Making a pie graph is more complex. Each pie wedge in a pie graph should represent an area equal to the percent or fraction of the whole. How do you compute the area of a pie wedge? You don't. Instead, convert all the magnitudes or values from the table to fractions that add up to 1. There are 360 degrees in a circle. For each category, multiply the fraction by 360. This yields the number of degrees for each corresponding wedge.

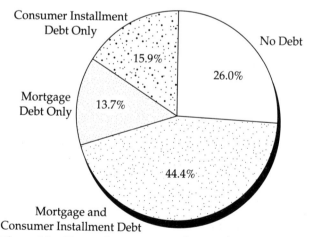

Figure 18.   Pie graph: mortgage and consumer debt outstanding for homeowners for 1986. *Source:* Bureau of the Census, *Statistical Abstract Of The United States, 1989* (Washington D.C.: GPO, 1989) 499.

Table 21.3 Preparing Values for a Pie Graph

Beverage	Original Value	Fraction	Number of Degrees
Milk	26.3	.1924	69.3
Tea	7.0	.0512	18.4
Coffee	26.7	.1953	70.3
Soft drinks	30.3	.2217	79.8
Juices	5.8	.0424	15.3
Beer	34.7	.2538	91.4
Wine	3.5	.0256	9.2
Distilled spirits	2.4	.0176	6.3
Total =	136.7		

Let's make a pie graph from the data on beverage consumption in Table 21.1. First add the values together to get a total. Then divide each original value by that total to get a fraction. Multiply each fraction by 360 to get the number of degrees each value contributes to the pie graph. In Table 21.3, the values are transformed into the desired degrees.

Using a compass, draw a circle and mark the center. Then draw one radius line from the center to the circle's edge. Using a protractor, (the little plastic tool that you can place over an angle to measure the number of degrees in the angle) measure an angle of 69.3 degrees

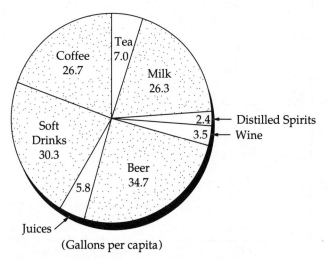

(Gallons per capita)

Figure 19. Pie graph: per capital consumption of beverages, by type for 1986. *Source:* Bureau of the Census, *Statistical Abstract Of The United States, 1990* (Washington D.C.: GPO, 1990) 125.

and draw another radius. In actual practice, you generally have to round to the nearest degree, as trying to measure 69.3 degrees is too difficult with a small home protractor. When you have drawn the 69-degree wedge, you have generated a pie wedge that represents 26.3 gallons of milk out of a total of 136.7 gallons of beverages. Continue measuring the angles for each wedge. Remember that half of a circle is 180 degrees, a quarter is 90 degrees, and an eighth of a circle is 45 degrees. When finished, you have Figure 19.

LINE GRAPHS

The line graph is in some ways similar to the vertical bar graph. The line graph shows change over time; points are placed at locations appropriate to the scale and then are connected.

Figure 20 is a line graph that shows the median earnings of year-round full-time workers by sex for the years 1960 through 1987. The

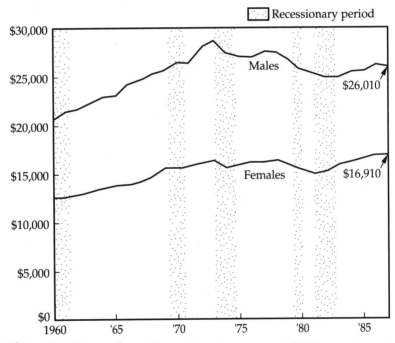

Figure 20. Line graph: median earings of year-round, full-time workers: 1960 to 1987 by sex (In 1987 dollars). *Source:* Bureau of the Census, *Population Profile of the United States, 1989,* (Washington D.C.: GPO, 1989) 33.

appropriate points were used to draw the graph but eliminated in the final version to give the line a more even look. The graph shows that women earn considerably less money than men do and that the gap is not closing appreciably.

SPECIAL GRAPHS AND CHARTS

Charts are frequently used to demonstrate processes rather than magnitudes. A common chart is the flowchart, which is used extensively in computer programming and scientific management. Figure 21 is a flowchart that shows how calls for emergency police service are received and handled in a modern police-communications center. Figure 22 demonstrates the operation of a computer program.

LYING WITH GRAPHS

You may have heard the old adage that liars figure and figures lie. This saying also applies to graphs. There are several ways that graphers can manipulate data from a table into a graph that misrepresents the truth. Sometimes this happens accidentally through ignorance, but at other times it is accomplished deliberately. It is important to recognize when such manipulation is taking place. Remember that graphs are pictures of data and are used in most cases to compare magnitudes. But one of the unexpected enjoyments you can derive from understanding graphs and charts is to catch others trying to cheat with them. In this section you'll learn two subtle techniques graphers use.

Let's look at an example from Volume 2 of *The Condition of Education, 1989*, by the National Center for Educational Statistics. Figure 23 shows the average salaries for college and university faculty from 1972 to 1986. The graphs compare salaries for different ranks of professors at public and private schools. What exactly do the two graphs show? It is obvious that salaries decreased between 1972 and 1981 and then began to increase. However, the increase has not caught up to the 1972 salaries. It also appears that the professors' salaries were drastically higher than either associate professors' or assistant professors' salaries. The graphs visually suggest that the poor assistant professors earned only a small fraction of what full professors earned and that, by 1981, their salaries almost dried up! Surely, the assistant professors were near the poverty level, whereas the full professors were raking it in.

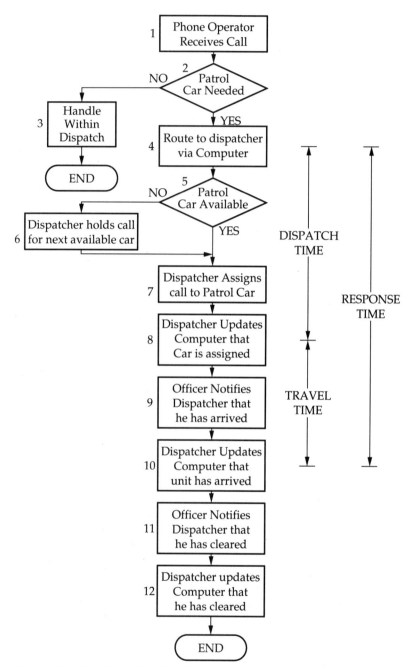

Figure 21. Flowchart of police dispatch operation

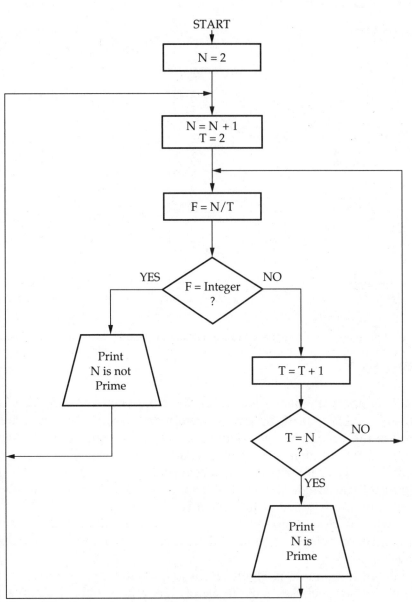

START

N = 2

N = N + 1
T = 2

F = N/T

F = Integer
?

YES

NO

Print
N is not
Prime

T = T + 1

T = N
?

NO

YES

Print
N is
Prime

Figure 22.   Computer program flowchart for finding prime numbers

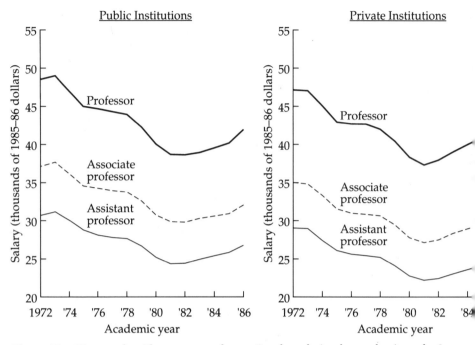

Figure 23.   Line graph with a suppressed zero. Faculty salaries, by academic rank. *Source:* National Center for Education Statistics, *The Condition of Education: 1989*, vol. 2 (Washington D.C.: GPO, 1989), 37.

Check the scale at the left. Aha! The scale begins not at 0 but at 20—that's $20,000. If the graphs were redrawn with 0 as the lower limit of the scale, you'd get Figure 24. Notice that, even though there are substantial differences among salaries, the plight of the assistant and associate professors is nowhere near as serious as the graph in Figure 23 suggested. This trick of beginning the scale with a number greater than 0 is called a suppressed zero.

If you pay close attention to graphs used in popular magazines and presented on television, you will frequently encounter graphs that have suppressed zeros.

Our next example is Figure 25, a graph showing the increase in public money spent per child in elementary and secondary education per year from 1970 to 1987. Instead of using two bars on a bar graph, I used two schoolhouses for the comparison. The expenditures per child increased between 1970 and 1987 (in constant 1986-1987 dollars) from $2,403 per child per year to $3,977 per child per year. This is

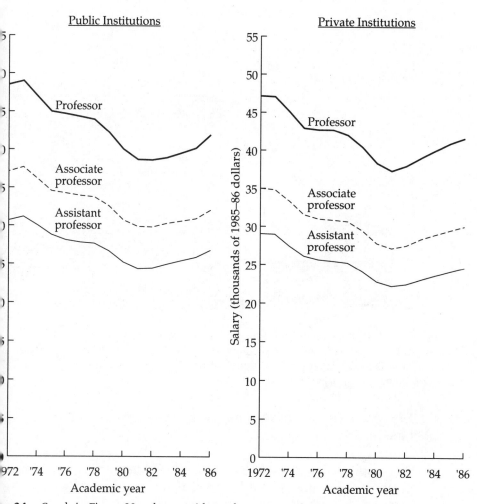

Figure 24.   Graph in Figure 23 redrawn without the suppressed zero.

a 65.5% increase. Now, look at Figure 25. The heights of the two flags on the schoolhouses correspond to the correct figures for the two years, but the graph gives the impression of a much greater increase. Why? The larger house has increased not only in height; it has increased also in breadth. The area of the large house is 170% greater than the area of the small house. The eye records this greater increase to leave the reader with the impression that the increase in public-school cost per child was really much greater than the figures indicate. Whenever you see a graph that is based on a vertical and horizontal increase in size, be careful. Someone is trying to trick you.

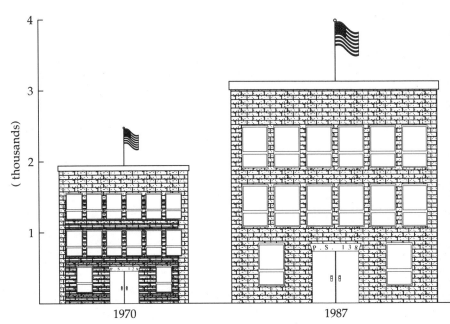

Figure 25. Graph illustrating disproportionate increase in size: current yearly expenditure per pupil in public elementary and secondary schools for 1970 and 1987. Data from National Center for Education Statistics, *The Condition of Education: 1989*, vol 1 (Washington D.C.: GPO, 1989) Table 1:14-2.

Let's recap the two most common types of errors (or deceptions) that are made with graphs.

1. The suppressed zero: Rarely is a suppressed zero called for. If you encounter one, ask why it is there. If the reason is not obvious, disregard the graph.
2. Fancy pictures: Is the graph made from a cute set of symbols (houses, boats), which change in breadth as well as height, thus giving the impression of greater increase or decrease?

WHAT HAVE YOU LEARNED?

1. Graphs and charts are simple pictures of data, relationships, or processes. Graphs should be simple to interpret.
2. The graph consists of categories, a scale, and a table of associated data. The vertical bar graph has the scale in a vertical position, usu-

ally at the left, and the categories along the bottom. The horizontal bar graph places the scale along the bottom and the categories at the left.

3. Pie graphs show magnitudes as wedge-shaped parts of a circle. To construct a pie graph, you must change the tabular data into degrees.

4. Watch for deceit in graphs, especially in the suppressed zero and in pictures that increase in breadth as well as height.

Of course, the point behind all these graphs is that they were built from statistics, which is exactly the subject of the next chapter. Come on; it's not that frightening.

CHAPTER 22

STATISTICS—MORE
PERSONAL POWER

You probably saw the word *statistics* in the chapter title and thought, "Oh, no! I think I'm going to be sick." Well, you don't have to feel queasy, because you're going to deal only with beginning statistics.

Statistics are an integral part of technology and science. They are also useful on an individual level. With the beginning statistics in this chapter, you will learn techniques useful in analyzing everyday problems. Wipe the sweat from your palms, and prepare to increase your personal power.

STATISTICS—WHAT ARE THEY?

Statistics make up a mathematical system for describing groups of things. Sometimes you have a group of events, or people, or objects, and you want to describe that group with a few numbers. At other times, you may have a group so large that you can't measure everyone or everything in the group; so you take a few samples of the group and measure them to get some idea about the group as a whole. In both cases, you are using statistics.

To avoid confusion, let's adopt some of the vocabulary used by statisticians. When you are studying a group of things, that group is called a *population*. Usually, a population means a group of people, but in statistics it has a much broader usage. In statistics a population is any collection of things you wish to measure and study.

But what is it you want to study concerning the chosen population? You want to study a characteristic or attribute. If your population is the collection of all third-grade students at your local public school, the characteristic of interest might be height, hair color, or number of siblings. After selecting a characteristic of interest, you measure that characteristic for some or all of the population (depending on how big the population is). The collection of measurements is called a *distribution*. The distribution is what you learn to work with in statistics. You will learn to calculate two different statistics: the mean and the median.

AVERAGE IS THE MEAN

The most common statistical measure is the average, which statisticians call the *mean*. You are probably so used to using the word *average* that you don't realize you're talking about statistics. Let's return to our third-grade class. You have 7 students. These 7 students make up the population. You decide to measure their heights and get the following results: 45.4 inches, 47.9 inches, 48.1 inches, 49.2 inches, 50.5 inches, 51.4 inches, and 52.6 inches. These 7 measurements become the distribution. What is the average height of the 7 students? Just add the 7 measurements and divide by the number of measurements.

The sum of the measurements is 345.1 inches. Divide this by 7 to get 49.3 inches. The average height of the 7 students is 49.3 inches, even though no particular student is exactly this height.

Definition
The mean (or average) is the sum of the measurements divided by the number of measurements.

Averages are easy to calculate and fun to apply to your own life situation. You use averages whenever you compute your car's average fuel consumption, average monthly electric bill, or average monthly food expense. Computing averages can help you to understand your budget problems.

However, there are situations in which the average either does not make sense or is a poor measure of a distribution. Suppose you have 20 teams of dogs and dog handlers at a dog show. Considering both the dogs and their trainers, what is the average number of legs per team member of the 40 individuals in the 20 teams? Well, it's 3 legs!

But that doesn't make any sense, because no one individual in the population has 3 legs. They have either 2 legs (trainers) or 4 legs (dogs).

Here's another example. What is the average gestation period in days for elephants and squirrels? This average turns out to be 334 days. But this number is nowhere near the average gestation period of either the elephant (624 days) or the squirrel (44 days). The average of 334 days is really a worthless statistic.

You can see what is wrong with the two preceding populations. They mixed apples and oranges. A population made up of obviously different subgroups can often yield misleading statistics.

Sometimes you run into problems with the average even when the population consists of the same kind of item. Let's go back to the 7 third-graders and suppose that the tallest student is not 52.6 inches but 75 inches tall (a very fast-growing child). Here's the new average.

$$\text{Mean} = \frac{367.5}{7} = 52.5 \text{ inches}$$

The new average is greater than 6 of the 7 measurements. You had one measurement that was not characteristic of the population (an outlier), and it threw the average off. This is always a danger in computing averages, and you must be aware that odd figures in some populations can distort the value of the mean.

PROBLEMS

Compute the average (mean) for each hypothetical distribution.

1. The length of trout caught by fishermen in a local lake as reported by the fishermen themselves: 12 inches, 14 inches, 17.5 inches, 25 inches, and 36 inches.

2. The actual length of the trout caught in Problem 1: 4.5 inches, 6.2 inches, 7 inches, 8 inches, and 11 inches.

3. The top speed of grizzly bears as reported by forest rangers: 28.5 mph, 29.9 mph, 31.1 mph, 32.6 mph, 33.0 mph, and 33.2 mph.

4. The top speed of the forest rangers: 24.1 mph, 25.8 mph, 26.3 mph, 26.4 mph, 27.1 mph, and 27.2 mph.

ANSWERS

1. Average length of trout caught as reported by the fishermen.

$$\text{Mean} = \frac{(12 + 14 + 17.5 + 25 + 36)}{5} = 20.9 \text{ inches}$$

2. Actual average length of trout caught.

$$\text{Mean} = \frac{(4.5 + 6.2 + 7 + 8 + 11)}{5} = 7.34 \text{ inches}$$

3. Average top speed of grizzly bears.

$$\text{Mean} = \frac{(28.5 + 29.9 + 31.1 + 32.6 + 33.0 + 33.2)}{6} = 31.4 \text{ mph}$$

4. Average top speed of forest rangers.

$$\text{Mean} = \frac{(24.1 + 25.8 + 26.3 + 26.4 + 27.1 + 27.2)}{6} = 26.15 \text{ mph}$$

COMPUTING THE MEDIAN

When the average is not a reliable statistic to use in measuring a distribution, you can compute what statisticians call the *median.*

Definition
The median is a measure such that half the other measurements are larger than the median, and half the measurements are smaller than the median.

From the definition, you can see that the median is a kind of halfway mark. To illustrate the median, let's consider the 7 third-graders and their heights.

45.4 inches

47.9 inches

48.1 inches

49.2 inches

50.5 inches

51.4 inches

52.6 inches

The various heights are in order from least to greatest. Count down the list to the middle measure, or 49.2 inches. You have found a number such that 3 (half) of the remaining 6 measurements are less than 49.2, and 3 (half) are greater than 49.2.

The median can be useful, for it can warn you when the average is not centered in the middle of the distribution but is off to one side. Consider the third-grade class after the 75-inch-tall student replaced the previously tallest student. The new distribution looked like this.

45.4 inches

47.9 inches

48.1 inches

49.2 inches

50.5 inches

51.4 inches

75.0 inches

In this new distribution the average, or mean, is 52.5 inches, which is greater than six of the seven measurements. When you compute the median for this distribution, you still get 49.2 inches. The median is not changed by a few measurements significantly different from the rest of the population. If the mean and the median are close together, the mean is probably a more useful statistic. If the mean and the median are far apart, it is generally better to rely on the median.

How do you compute a median when you have an even number of measurements? When you have an even number of measurements, simply select the two central measurements, add them, and divide by two. If the 75-inch third-grader were to leave school on a basketball scholarship, you would have these heights left.

45.4 inches

47.9 inches

48.1 inches

49.2 inches

50.5 inches

51.4 inches

The two central measurements are 48.1 and 49.2. Add them to get 97.3. Divide this by 2 to get 48.65 inches for the new median. Now try the sample problems.

PROBLEMS

Compute the median for each of the four distributions given in the sample problems on page 284.

ANSWERS

1. Since there are five terms, the middle one is the median: 17.5 inches.

2. Again, there is an odd number of terms; the median is 7 inches.

3. Here you have an even number of measurements. Select the two central terms, 31.1 and 32.6, add them, and divide by 2 to get the median, 31.85 mph.

4. As in Problem 3, select the two central terms, 26.3 and 26.4. The median is 26.35 mph.

WHAT CAN GO WRONG?

You have explored two statistical measures that you can use in your own life. In all likelihood, you will never need to compute more complicated statistics or scientific testing for your personal use. However, you may at some point be the victim of faulty or deliberately biased statistics compiled by others. Most statistical tests carried out in the scientific community are done according to sound procedures and theory. When mistakes are made, you generally don't hear about them.

The kind of statistics you are confronted with daily are from advertising agencies and politicians who bombard you with data to persuade you to try their products or give them your vote. In almost every case, you should ignore such numbers. Because of errors, bias, and special interests, the chances are good that the data will be meaningless. Actually, you can have fun trying to catch flimflam artists in the act. Knowing the fallacies that are possible in statistics increases your enjoyment of reviewing data and understanding arguments.

Here's a short list of some of the mistakes, errors, or frauds that you may encounter.

1. *Special interests.* Ask yourself if those presenting statistical data have a special interest to promote. If so, you must be doubly suspicious of their claims. Even a professionally trained scientist has difficulty excluding bias from his or her testing procedures. Nonprofessionals often deliberately manipulate data to show their cause in a favorable light.

2. *Meaningless claims.* The entire argument supporting the claim may be meaningless because it cannot be tested. A fortune teller says the world is coming to an end because an ancient warrior spirit spoke to her in a dream. How can this prediction be tested before the event? It can't. A claim must also be useful. A statistician claims that changes in the weather control world oil prices. Even if true, this knowledge is virtually useless since you can't accurately forecast the weather.

3. *Additional test results.* When someone reports test results to you, are they giving you the results of all the tests, or of only those tests that put their claim in a favorable light? A product can be tested many times, but only a fraction of the tests—those with favorable outcomes—are provided to the public.

4. *Sample bias.* Sometimes you cannot measure an entire population, so you have to measure a sample or a subset of the population. If you choose your sample improperly, you introduce bias into the distribution. Adding bias by sampling selectively is a trick used often by those who want to lie with statistics. You can prove almost anything if you are clever in selecting your sample. Valid samples should be random, which means that every member in the population should have the same chance of being selected for the sample.

5. *Small sample.* A small sample can generate as much error as a biased sample. Some advertising claims are based on small samples, which have a greater margin of error, and can therefore be inaccurate.

6. *Mean instead of median.* For some populations, the mean is often deceptive, and the median is a better statistic. If the mean is used in an argument, how can you be sure it's the appropriate statistic?

7. *Vague test details.* Honest researchers are proud of their work and will give you all the details you want, including number of sam-

ples, sample sizes, how the samples were drawn, and how the results were tabulated. Charlatans will not reveal such details because their tests don't hold up under scrutiny. One TV advertisement goes something like this. "*Some* recent government studies have *suggested* that the risk of *some* forms of cancer *may* be reduced with a *high* bran diet." Balderdash! I have italicized the key words to dramatize the ambiguity in such a claim. How much better it would be to state, "Ten government studies conducted during 1989 prove that the risk of three kinds of cancers is reduced by 50% with a diet 20% higher in bran." The ad agency didn't provide the second kind of statement because it couldn't—it didn't have those statistics.

8. *Invalid argument form.* Sometimes the logic of the argument is twisted and misused. Watch for any of the following invalid arguments.

   A. *Post Hoc.* This logical fallacy is simply that, if A occurs before or during B, then it must have caused B. Examples of this argument form are abundant: Harry Truman was president at the outbreak, of the Korean War, and therefore caused the war; increasing the national debt has driven the stock market up, and ocean pollution has caused the collapse of socialism in Central Europe. Frequently, this argument form is used to connect two phenomena that people think are already connected. This enhances the attractiveness of the argument: Prison overpopulation is caused by violence on TV; increased drug use is the result of increased permissiveness. Whether two phenomena are connected is not demonstrated by simply showing that a change in one preceded or accompanied a change in the other.

   B. *Argument from authority.* "Dr. Jones from MIT says the universe is collapsing, and he should know." Anytime an argument leans on the credentials of an expert, you must be careful. Sometimes people don't even care if the expert's field is related to the field under discussion. Would it make sense to have beautiful young movie stars testify before Congress about farm foreclosures? Do they know something about farming that a farmer doesn't?

   C. *Argument against the person.* If you can destroy the character of an individual, you destroy his or her argument. This technique is used frequently in political campaigns. While good character is a desirable trait in any person, the argument one offers should be considered on its own merits.

It's more fun to listen to arguments and review data when you know the kinds of errors that can be made. Catching someone who is trying to mislead you makes you feel in control and puts you in charge of your own destiny.

WHAT HAVE YOU LEARNED?

Statistics are a powerful tool in industry and science. They can also be useful in your private life. You have learned several new terms and two useful statistical measures.

1. The collection of items you are studying is called a population.
2. That feature of the population under study is the characteristic or attribute.
3. A collection of measures of an attribute is a distribution. From the distribution, you can calculate the mean (average) and the median.
4. Definition: The *mean* (or average) is the sum of the measurements divided by the number of measurements.
5. Definition: The *median* is a measure such that half the other measurements are larger and half are smaller.

You have also learned to watch for some dangerous errors or frauds.

1. Are there special interests involved?
2. Is the claim under consideration meaningful?
3. Do additional test results exist ?
4. Are the samples biased?
5. Are the samples large enough?
6. Was the mean used where the median was appropriate?
7. Have test details been made available?
8. Has an invalid argument been used?

You deserve a reward for having come this far. There's one more chapter, but it consists of puzzles and amusements. For someone who used to be a mathophobe, you've come a long way. Aunt Elsa and Uncle Ozzie are very proud of you. (And that niece of yours is probably envious!) On to the finish line!

CHAPTER 23

PUZZLES AND

PLEASURES

The driving energy of math consists of wonderment and the joy of discovery. Math is something you first play with; only after the play does it become practical. In this chapter you will be presented with puzzles and curiosities of mathematics. They are not offered to help you solve problems, but for entertainment.

MAGIC SQUARES

You have already been exposed to numerous math games including Monopoly, blackjack, and various dice games. Presented here is a kind of game based on an ancient pastime involving magic squares.

A magic square is a large square that has been subdivided into a number of smaller squares or boxes. Numbers are placed in the boxes so that the rows, columns, and two major diagonals all add to the same number. Figure 26 is an example of the earliest known magic square, which comes from China and dates back to 2200 B.C. Notice that the nine boxes have been filled with the digits 1 through 9 in such a way that the three rows, three columns, and two diagonals all add to 15.

There are many different magic squares. Magic squares can be constructed of 9, 16, 25, or more boxes. The basic square in Figure 26 can be used to make many more squares consisting of 9 numbers. Before learning how to make different 9-box squares, let's look at how the magic square in Figure 26 was completed.

Figure 27 shows our square plus eight additional squares attached to the four sides and four corners. Begin by placing the 1 in the top middle box of the center square. Now you use two simple rules to place the rest of the numbers.

8	1	6
3	5	7
4	9	2

Figure 26. Magic square

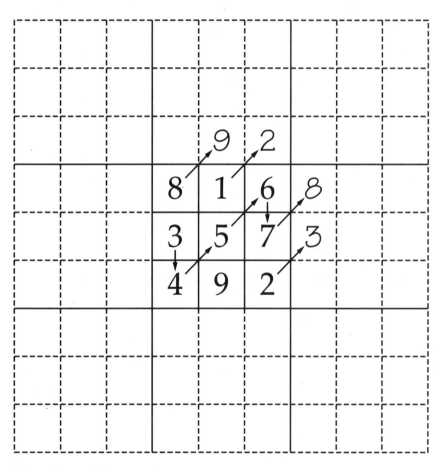

Figure 27. Building a magic square

Rule 1

Each successive number is placed diagonally up and to the right. If that move takes you into one of the adjacent squares, place the number in the corresponding box in the center square.

Rule 2

If a box is already occupied, place the number in the box below its predecessor.

Notice in Figure 27 that the next number, 2, has been placed in the lower right-hand-corner box of the adjacent square. This corresponds to the lower right-hand box in the main square, so that is where to place it. Now the 3 is placed diagonally up and to the right, according to Rule 1. This puts it into the center box of the left column of the adjacent square. This corresponds to the center box of the left column of the main square, so place the 3 there. The 4 should go in the box diagonally up and to the right of the 3 box, but that box already has 1 in it. Using Rule 2, place the 4 in the box below the 3.

Next, the 5 goes diagonally up and to the right of the 4, followed by the 6 in the upper-right box. You can't use Rule 1 to place the 7 in the box diagonally up and to the right, so use Rule 2, and place the 7 in the box below the 6. Using the two rules, place the numbers 8 and 9.

If you remember the two rules, and begin with 1 in the middle box of the top row, you can always reconstruct the basic magic square.

Now the fun begins, because you can use your knowledge of the basic 9-box square to build other 9-box squares. What happens when you add the number 1 to all 9 boxes in the magic square? You get another magic square in which the combinations of boxes add to 18 instead of 15. In fact, you can add or subtract any number to all 9 boxes and generate additional magic squares.

Can you multiply or divide all 9 boxes by the same number and still get another magic square? Can you design a magic square in which the boxes all add to the number 1 or 0? Be careful—playing with magic squares is addictive.

You can use knowledge of magic squares to dazzle your friends. Draw an empty 9-box magic square. Ask a friend to select a number and place it in one of the boxes. Then, based on the rules for constructing a magic square, fill in the remaining 8 boxes. It takes a little practice to learn the rules, but it's well worth it.

FIBONACCI NUMBERS

The Fibonacci sequence of numbers is a sequence that begins with the number 1. Each additional term is the sum of the two previous terms. The second number is the sum of 1 (the first term) plus the previous term, which does not exist. Therefore, the second term is just 1 again. The third term is the sum of the first two terms, 1 and 1, which is 2. The next term is the sum of 2 and 1, or 3. These are the first 15 terms of the Fibonacci sequence.

1, 1, 2, 3, 5, 8, 13, 21, 34, 55, 89, 144, 233, 377, 610, . . .

To form each term, just add the two previous terms. So why is there such a big deal about this sequence?

As it turns out, the Fibonacci sequence of numbers appears many times in nature in relation to how plants and animals grow and reproduce. The construction of many plants and flowers imitates Fibonacci numbers.

Now let's try some other sequences. Look at the numbers given, and try to figure out the pattern so that you can guess the next number in the sequence: 1, 2, 4, 8, 16, 32 . . . The next number is 64, since each number is 2 × the previous number. Here are two more patterns. Try to figure out the next number in each sequence.

Pattern 1: 1, 3, 7, 15, 31 . . . The next number will be 63; the pattern is 2 × the previous number, + 1 (2 × 31 = 62 + 1 = 63).

Pattern 2: 1, 2, 6, 24, 120 . . . The next number will be 720; the pattern is to multiply the first number by 2 to get the second number, the second by 3 to get the third number, and so on: 6 × 120 = 720.

You might want to try inventing some patterns of your own. Try them out on some friends—it's an amusing way to pass the time on rainy days.

The kinds of relationships that tie together two seemingly divergent fields of mathematics are common. The Fibonacci sequence was discovered by a mathematician playing with numbers and seeing how they could form a sequence. This demonstrates why play is so important to math. You, too, can fill the hours with math play.

WHERE DO YOU GO NEXT?

Congratulations on successfully completing this book. You are now more knowledgeable in mathematics than your friends or relatives (unless you are in the habit of running around with mathematicians). You are certainly more knowledgeable than the majority of people who have lived before you and the mass of people who occupy the planet today. But why stop now?

Exactly where are you on the mathematics staircase? In terms of formal study, you are now proficient in arithmetic. You've been introduced to algebra and can solve simple equations with one unknown. To increase your proficiency, you need more algebra and an introduction to geometry. Next comes analytic geometry, which can be followed by calculus. At any time, feel free to study number theory. Several important discoveries in number theory have been made by amateur mathematicians.

Nothing can prevent you from studying any of these subjects from a nontechnical point of view. You can proceed with two types of books: the self-help book (much like this one), which will provide you with technical information, or the kind of book that talks about math but does not actually teach math skills. Both kinds of books are necessary. The bibliography has an annotated list of books that can help you.

Don't be limited to the list in the bibliography. When you are browsing in a bookstore, make it a habit to wander to the math section and review the volumes displayed. You will be pleased with the wide range of math books that are being written for the general public. Also, check your public library. Bookstores that sell used books often undervalue their math books, and you can pick them up for a fraction of their real worth.

Regardless of whether you ever pick up another book about mathematics, you should continue to integrate math into your everyday life. You should continue to check store receipts, plan your budget, and make savings investments. You will keep control of your life where it should be—in your hands. However, if something you have read in this volume has sparked your interest and curiosity, then by all means, read and study more about math. As with any endeavor, the more proficient you become, the more you will enjoy it.

Up to now, you have taken the responsibility of curing your math phobia and providing for yourself a sound foundation in mathematics. But if you have children, your responsibility goes beyond yourself. You live in a country supported by technical advancement and achievement. The future will demand more engineers, scientists, and mathematicians. Encourage your children to take mathematics and science courses. Show them that you read math books, and encourage them to talk about their math lessons. Enjoy math puzzles and games together.

I have enjoyed our time together. If you have learned about mathematics and enjoyed some historical insight into its roots, then I am amply repaid for my efforts in writing this book. Perhaps, someday, you and I will bump into each other in the back of a bookstore, in the section labeled *MATHEMATICS*. Nothing could be better proof that you have conquered your math phobia once and for all.

GLOSSARY

abacus a device used from ancient times to the present for performing arithmetic operations.

addition the fundamental operation of arithmetic; combines two or more numbers into a sum.

algebra the study of the general laws of arithmetic, including the solving of equations containing unknowns.

amortization table a table that shows the interest, reduction in principal, and remaining balance for various payment periods on a mortgage.

analytical geometry the branch of mathematical study that combines both algebra and geometry.

approximation the technique of quickly estimating answers.

arithmetic the first field of mathematical study; covers the four basic operations: addition, subtraction, multiplication, and division.

associative law of addition if A, B, and C and numbers, then $A + (B + C) = (A + B) + C$.

associative law of multiplication if A, B, and C are numbers, then $A \times (B \times C) = (A \times B) \times C$.

bar graph   a graph that represents each value as a single bar.

borrowing numbers   in long subtraction, the process of subtracting 1 from the left column and adding 10 to the current column.

bringing down a number   in division, the process of transposing a number from the number being divided in order to continue the division process.

calculus   the study of measurement through the use of limits.

cancellation   the elimination of the same number from both the top number and bottom number of a fraction.

carrying numbers   in long addition and long multiplication, the process of moving the left digit of a two-digit sum or product to the next column on the left.

category   in graphing, the definition of a specific value to be graphed.

chart   a visual presentation showing relationships between items in a process.

circumference of a circle   the length of the line that defines a circle.

circle   the collection of points equidistant from a fixed point; whose area is equal to πr^2, where r is the radius.

commutative law of addition   if A and B are numbers, then $A + B = B + A$.

commutative law of multiplication   if A and B are numbers, then $A \times B = B \times A$.

compound interest   a type of interest charged against both the principal and outstanding interest of a loan, whose formula is: $S = P(1 + R)^n$ where S is the maturity value, P is the principal, R is the interest rate, and n is the number of compounding periods.

cross multiplication   the multiplication of both the top number and bottom number of each of two fractions by the bottom number of the other fraction.

cube   an enclosed volume of six sides, with each side a square; whose volume is equal to (length of a side)^3.

cube root   a second number that, when multiplied by itself three times yields the original number.

cylinder   an enclosed tubelike volume with a circle on the bottom and top and one curved vertical wall; whose volume is equal to $(\pi)(\text{radius})^2 \times$ length.

decimal   a number written in the base 10 system, in which each digit is ten times the same digit located to the immediate right.

decimal point   a point used in a decimal to lock the position of the digits; located to the immediate right of the units digit.

diameter of a circle   the length of a straight line drawn from the circumference of a circle through the center of the circle and to the circumference again; a line that evenly divides a circle.

digit   one of ten symbols, 0 through 9, used in the modern Hindu-Arabic numbering system.

distribution   in statistics, a collection of measurements gathered from a population.

distributive law   if A, B, and C are numbers, then $A \times (B + C) = A \times B + A \times C$.

division   one of the four basic operations of arithmetic, defined as the reverse of multiplication.

equal sign   a symbol used to show equality between two numbers or combinations of numbers.

equation   the symbolic representation of two combinations of numbers as equal through the use of an equal sign.

even number   any whole number that can be divided by 2 without a remainder; any whole number that has 2 as one of its primes.

exponent   a positive whole number written as a superscript to a second number (the base), which is shorthand for the number of times the second number should be multiplied by itself.

factor   to separate a whole number into its prime numbers.

Fibonacci sequence   a sequence of numbers beginning with 1 where each number is the sum of the preceding two numbers, i.e. 1, 1, 2, 3, 5, 8, 13, 21, ...

finite   a quantity that is limited and can be represented by a number.

formula   a rule for computing a desired quantity written in algebraic form.

fraction   a number written as one whole number divided by a second whole number where neither number is 0.

fraction in lowest terms   a fraction whose top number and bottom number do not share a common prime.

geometry   the study of shape and forms.

graph   a picture of tabular data.

Hindu-Arabic number system   our current positional number system based on the ten digits, 0 through 9.

horizontal bar graph   a bar graph whose scale is on the bottom and categories are on the left.

inequality   the relationship where two quantities are not equal.

infinite   without bounds; as a quantity, not represented by any real number.

infinite decimal   a decimal that contains an infinite number of digits.

infinite, repeating decimal   an infinite decimal containing a digit sequence that repeats itself indefinitely.

interest   the money charged for borrowing money.

interest rate   a percentage or decimal that, when multiplied by the principal, yields the interest owed.

law of equal addition and subtraction   if $A = B$, then $A + C = B + C$ and $A - C = B - C$.

law of equal multiplication and division   if $A = B$, then $A \times C = B \times C$ and $A/C = B/C$ where C is not 0.

law of substitution   equals can be substituted for equals.

line graph   a graph of connected dots where the value of each category is represented by one of the dots.

long addition   the process for adding whole numbers or decimals by stacking the numbers and applying the principle of carrying.

long division   the division algorithm allowing for the division of any two whole numbers or decimals by applying the principle of bringing down a number from the number being divided.

long multiplication   the process of multiplying two whole numbers or decimals by stacking and applying the principle of carrying.

long subtraction   the process of subtracting two whole numbers or decimals by stacking and applying the principle of borrowing.

magic square   a square subdivided into boxes where each box contains a number so that the rows, columns, and diagonals all add up to the same number.

mathematics   the study of numbers and magnitudes.

maturity value   the sum of the interest and principal on a loan that is due at the maturity of the loan.

mean   in statistics, the sum of the values divided by the number of values.

median   in statistics, a value such that half the other values are larger and half are smaller.

minus sign   the symbol ($-$) used to indicate the operation of subtraction and to identify negative numbers.

multiplication   one of the four basic operations in arithmetic; defined as repeated addition.

negative number   any number to the left of 0 on the number line.

number   any entity that when operated on by the laws of mathematics does not violate those laws.

number line   an infinite straight line with a point representing 0, with negative numbers marked off to the left of 0 and positive numbers marked off to the right of 0.

number sequence   a set or collection of numbers that can be either finite or infinite.

number theory   the study of whole numbers.

odd number   a whole number that, when divided by 2, leaves 1 as a remainder; any whole number lacking 2 as one of its primes.

pebble number   a method used by ancient Greeks for writing whole numbers.

percent sign   the symbol (%) written on the right of a number identifying the number as a percentage.

percentage a number representing parts per hundred; a fraction whose bottom number is understood to be 100.

pi the ratio of the circumference of a circle to its diameter, represented as π.

pie graph a graph constructed as a circle with pie-shaped wedges representing the different categories.

plus sign the symbol (+) used to indicate the operation of addition and also to identify positive numbers.

population in statistics, the set of items which share a common characteristic and are selected to be studied by statistical methods.

positive number any number to the right of 0 on the number line.

power the resulting value of a number with an exponent.

prime number a whole number that can be divided evenly only by the number 1 and by itself.

principal the original amount of a loan.

pyramid number any whole number that in pebble notation can be written in the shape of a triangle; the sum of consecutive integers beginning with the number 1.

radius of a circle the distance from the circumference of a circle to the circle's center.

radix a symbol representing the root of a number.

random sample in statistics, a sample where every member of the population has the same chance of being selected in the sample.

rectangle a closed, four-sided figure with equal interior angles of 90 degrees, whose area is equal to (length)$\times$(width).

rectangular volume a volume enclosed by three pairs of rectangles; whose volume is equal to (length)$\times$(width)$\times$ (depth).

reflexive law   if A is a number, then $A = A$.

remainder   the fraction of a number left over from division.

right triangle   a triangle with one interior angle of 90 degrees, whose area is equal to $(1/2) \times$ (length) $\times$ (height).

root of a number   the nth root of a number is a second number that when multiplied by itself n times is equal to the original number.

rounding a number   to decrease the number of digits in such a way as to minimize error.

sample   in statistics, a subset of a population selected for measuring some attribute of the population.

scale   in graphing, the range of values, beginning with 0, that the various categories can take.

short addition   addition directly from memory or finger counting.

short division   division directly from memory or where the number doing the division is a single-digit number.

short multiplication   multiplication directly from memory.

short subtraction   subtraction directly from memory or finger counting.

similar form   two fractions are in similar form when they have the same number for their bottom numbers.

simple interest   interest charged only against the principal of the loan during the course of the loan, whose formula is: $S = P(1 + RT)$ where S is the loan's maturity value, P is the principal, R is the interest rate, and T is the duration of the loan.

simplification   in algebra, to manipulate an equation until the unknown is reduced to a single number.

simplifying a fraction   reducing a fraction to its lowest terms.

sphere   the collection of points in three dimensions that are all the same distance from a fixed point; whose volume is equal to $(4/3)(\pi) \times R^3$.

square   a closed figure with four equal sides and four interior angles of 90 degrees; whose area is equal to (length of one side)^2.

square number   any whole number that in pebble notation can be written in the shape of a square; any whole number with a whole number square root.

square root   the square root of a number is a second number that when multiplied by itself is equal to the original number.

squaring a number   multiplying a number by itself.

statistics   the means of studying characteristics of populations through measurement and sampling.

subtraction   one of the four basic operations of arithmetic; defined as the addition of a negative number.

suppressed zero   in graphing, beginning the scale with a number greater than 0 to give the reader a false impression that the differences in magnitudes are greater than they are.

table of values   in graphing, the data used to make a graph.

tens place   in the modern numbering system, the position to the left of the units place.

terminating decimal   a decimal that contains only a finite number of non-zero digits.

triangle   a closed figure with three sides; whose area is equal to $(1/2) \times$ (height) $\times$ (base).

unit of measurement   a standard or scale used to measure with; e.g., foot, inch, degree, pound.

units place   in the modern numbering system, the position to the left of the decimal point for decimals, or the extreme right position for integers.

unknown   in algebra, a number being solved for; generally written as X or another letter.

vertical bar graph   a bar graph with categories along the bottom and the scale on the left and/or right side.

whole number   one of the numbers we learn to count with, beginning with 0 and increasing or decreasing in increments of 1.

BIBLIOGRAPHY

Beckman, Peter. *A History of* π. New York: St. Martin's Press, 1971. A delightful, irreverent book containing much more than the history of π. Definitely worth reading.

Bell, E.T. *Mathematics: Queen and Servant of Science.* New York: McGraw-Hill Book Company, 1951. A classic book that gives a comprehensive overview of mathematics; I highly recommend it.

Boyer, Carl B. *A History of Mathematics, Second Edition.* New York: John Wiley and Sons, 1968, 1990. The revised edition of a classic, detailed book that gives a very complete history of mathematics. The style is clear and the many examples enhance the readability. I highly recommend it.

Broudy, Rose L. *Modern Math Made Easy.* New York: Harvey House, 1970. A short volume dealing with the basic definitions of arithmetic, including an introduction to sets.

Bunt, Lucas N., Jones, Phillip S., & Bedient, Jack D. *The Historical Roots of Elementary Mathematics.* New York: The Dover Publications, 1988. This fine book deals with the development of mathematics through the ancient Greeks.

Cutler, Ann, & McShane, Rudolph. *The Trachtenberg Speed System of Basic Mathematics.* New York: Doubleday & Company, 1960. This book is for anyone who wants to develop the skill of rapidly performing basic mental arithmetic on very large numbers. The system works, but cheap, available calculators decrease the need for such a technique.

Dunham, William. *Journey Through Genius: The Great Theorems of Mathematics.* New York: John Wiley & Sons, Inc., 1990. This collection of 12 mathematical masterpieces spanning 2300 years is beautifully written for the lay reader. Each chapter tells a story, putting a historical and biographical context around the mathematician and his theorem. I strongly recommend this book.

Frohlichstein, Jack. *Mathematical Fun, Games and Puzzles.* New York: Dover Publications, 1962. This is an excellent book; it teaches basic math and math applications through projects, games, and puzzles.

Glick, James. *Chaos: Making a New Science.* New York: Penguin Press, 1987. This book is for the intellectually courageous, yet its rewards are great. The book details the development of the new science (math) of chaos theory with personal sketches of those responsible. The more you know of math, the more you'll enjoy this book. The pictures alone are worth the price.

Hardy, G.H. *A Course of Pure Mathematics.* Cambridge: The University Press, 1963. This is an advanced book on mathematics for that individual tired of basic math. It covers everything from functions of real variables and complex numbers to calculus. The notation is detailed and complete.

Heath, Sir Thomas. *A History of Greek Mathematics.* Oxford: The Clarendon Press, 1960. This is a classic history of Greek mathematics; very scholarly.

Helton, Floyd F. *Introducing Mathematics.* New York: John Wiley & Sons, 1958. A fine text that covers arithmetic, algebra, and geometry.

Hoffman, Paul. *Archimedes' Revenge.* New York: W.W. Norton & Company, 1988. This is a fun book to read, introducing different mathematical concepts with an entertaining, light style. Highly recommended.

Huff, Darrell. *How to Lie with Statistics.* New York: W.W. Norton & Company, 1954. This is an old book but well worth checking out of the library.

James, Glenn, & James, Robert C. (eds.) *Mathematics Dictionary.* New York: D. Van Nostrand Company, 1959. This is a fine mathematics dictionary, but somewhat dated. I recommend that you find yourself a good mathematics dictionary.

Kelly, Gerard W. *Short-Cut Math.* New York: Dover Publications, 1984. This volume presents short-cut methods for arithmetic. It is a good little book for sharpening your calculating skills.

Kline, Morris. *Mathematics and the Physical World*. New York: Thomas Y. Crowell Company, 1959. This book combines math and elementary physics to show how mathematics relates to the world around us.

Kogelman, Stanley, & Warren, Joseph. *Mind Over Math*. New York: McGraw-Hill Book Company, 1978. An interesting little book that reviews the psychological aspects of math phobia.

Madachy, Joseph S. *Mathematics on Vacation*. New York: Charles Scribner's Sons, 1966. This is a good book for games and entertainment, especially the section on magic squares.

Madigan, Bob, & Kasoff, Lawrence. *The First-Time Investor*. 1986. New York: Prentice Hall Press, 1986. A good book to review different investment options.

Mira, Julio A. *Arithmetic: Clear and Simple*. New York: Barnes & Noble, 1965. This is a fine book for reviewing and practicing basic arithmetic with lots of fun sample problems.

Moore, John T. *Fundamental Principles of Mathematics*. New York: Holt Rinehart and Winston, 1960. This is an advanced text that covers calculus, systems of equations, analytic geometry, and statistics.

Muir, Jane. *Of Men and Numbers*. New York: Dodd, Mead & Company, 1961. This is a highly recommended book that details the lives of great mathematicians. The book is great fun.

Newman, James R. (ed). *The World of Mathematics*, 4 vols. New York: Simon and Schuster, 1956. This is a classic four-volume collection of articles and can still be found in bookstores. Since each section was written by a different author, the quality varies greatly. However, many parts are worth reading and many give a unique slant on mathematics.

Pappas, Theoni. *The Joy of Mathematics*. California: Wide World Publishing/Tetra, 1989. This is a wonderful book that gives over 140 glimpses into math oddities and fun constructions. It is highly recommended for loads of enjoyment.

Paulos, John Allen. *Innumeracy*. New York: Hill and Wang, 1988. This excellent book defines math illiteracy and gives examples of its many consequences. Highly recommended.

Rucker, Rudy. *Infinity and the Mind.* New York: Bantam Books, 1982. This book is a great treasure of material on infinity. It also includes a layman's approach to understanding formal mathematical systems.

Selby, Peter H., & Slavin, Steve. *Practical Algebra, Second Edition.* New York: John Wiley & Sons, 1974, 1990. A wonderful book for reviewing and practicing basic arithmetic through algebra. It is full of many examples and sample problems.

Skane, Donna & Skane, Lawrence. *Mathematics for Consumers.* Massachusetts: Addison-Wesley Publishing Company, 1980. A good book for reviewing math as applied to personal finances.

Slavin, Steve. *All the Math You'll Ever Need.* New York: John Wiley & Sons 1989. An excellent book for learning to apply basic mathematics to everyday problems. Highly recommended.

Sloan, Robert W. *An Introduction to Modern Mathematics.* New Jersey Prentice-Hall, 1960. A slim volume that introduces basic concepts of sets mathematical logic, and the theory of functions.

Sticker, Henry. *How to Calculate Quickly.* New York: Dover Publications 1955. This is an exceptional book for practicing arithmetic, including mental arithmetic.

Tobias, Sheila. *Overcoming Math Anxiety.* Boston: Houghton Mifflin Company, 1978. This very readable book reviews many aspects of math phobia especially in relation to women. Highly recommended.

VanCaspel, Venita. *Money Dynamics for the New Economy.* New York: Simon and Schuster, 1986. A good beginning book for reviewing personal finance and investments.

von Mises, Richard. *Probability, Statistics and Truth.* New York: Dover Publications, 1981. This is a good book to get deeper into the subject of statistics. The text does not overpower the reader with equations and formula.

Weinberg, George H., & Schumaker, John A. *Statistics: An Intuitive Approach.* California: Wadsworth Publishing Company, 1962. This is one of the best books available for a solid introduction to statistics. It is easy to read with concepts introduced in an enjoyable, digestible manner. Very highly recommended.

INDEX

Abacus, 8, 267
Adding on, 200–202
Addition:
  breaking numbers down, 25–26, 36
  carrying numbers, 27, 29, 31–33, 268
  checking accuracy of, 30
  by columns, 34–35, 36–37
  commutative law of, 124, 125, 139, 269
  commutative property, 24
  decimals, 99–101, 113
  defined, 22, 38, 267
  as a definition of positive numbers, 22
  fingers, using to calculate, 23
  of fractions by cross-multiplication, 84–86, 97
  of fractions by simplification, 87–90, 97
  long, 27–29, 36, 271
  memorizing answers, 23
  by mental arithmetic, 200–207, 216
  number order, 23–24, 36
  of opposite sign negative numbers, 71–72, 76
  of same sign negative numbers, 71, 76
  short, 23–24, 36, 274
  shortcuts for, 23
  as a straight line, 22
  sum, 48

  zeros in, 24, 36, 203
Addition table for numbers under 10, 23, 26
Algebra, 131, 265, 267
Algebraic expression:
  for division, 132
  for multiplication, 131
Amortization table, 196, 267
Analytical geometry, 265, 267
Angle, 174
Applications of mathematics principles:
  distance problems, 170–171
  electrical usage, 164–165
  formulas, using, 163–164
  gas consumption, 167
  rounding numbers, 161–163
  unit of measurement, determining, 159–161
  unit pricing, 167–169
Approximation, 210–211, 267
Area, *see* Volume:
  of a circle, 153, 178, 185
  complex, 179–180, 185
  defined, 174
  of a rectangle, 175, 185
  of a square, 174–175, 185
  of a triangle, 175–177, 185
  as a unit of measurement, 159
Arguments (statistics), 259–260
Arithmetic, 4, 267. See also Addition; Division; Multiplication; Subtraction

errors/frauds, 257–260
invalid arguments, 259–260
mean, 253–254, 258, 260
median, 255–257, 258, 260
population, 252, 258, 260
uses of, 252
Sticker, Henry, 280
Stock market, 234, 235
Subtraction:
borrowing, 40, 42–44, 47,
268
checking accuracy of, 45
of decimals, 101–103, 113
defined, 38, 275
of fractions, 91–93
long, 40–45, 47, 271
by mental arithmetic, 208–209,
216–217
of negative numbers, 73–74
number order, 39, 47
short, 39–40, 47, 274
shortcuts for, 39
as a straight line, 38
Subtraction table, 39, 47
Sum, 48
Superscript, 151
Suppressed zero, 248–249,
275

Table of values, 237, 241, 250,
275
Tens place, 9, 11, 275
Terminating decimal, 17, 21, 110,
115, 117, 275
Thousands place, 11
Tobias, Sheila, 280
Triangle(s):
area of, 175–177, 185
defined, 175, 275
oblique, 176–177, 185
right, 175, 185, 275
Triangle number, 8

Unit of measurement:
area as, 159
defined, 275
determining, 159–161
Unit pricing, formula for, 167–169,
173
Units place, 276
Unknown, 131–138, 139, 164,
277
isolating, 133–138

VanCaspel, Venita, 280
Vertical bar graph, 237–240,
250–251, 276
Volume:
conversions, 168–169
of a cube, 156,181, 185
of a cylinder, 154, 183–184, 185
rectangular, 181–182, 185, 273
of a sphere, 154, 184, 185
von Mises, Richard, 280

Warren, Dr. Joseph, 2, 279
Weight conversions, 168
Weinberg, George, 280
Whole numbers, 7, 276

X, see Unknown

Zero(s):
in addition, 24, 36, 203
in decimals, 17, 108–109
in division, 63, 75, 108–109
as exponents, 152
in fractions, 14–15, 20
in the Hindu-Arabic number
system, 9, 11
in multiplication, 50, 52, 58, 75
suppressed, 248–249, 275